Machine Learning Applications in Medicine and Biology

Ammar Ahmed • Joseph Picone
Editors

Machine Learning Applications in Medicine and Biology

Springer

Editors
Ammar Ahmed
Advanced Safety and User Experience
Aptiv
Agoura Hills, CA, USA

Joseph Picone
Department of Electrical and Computer Engineering
Temple University
Philadelphia, PA, USA

ISBN 978-3-031-51892-8 ISBN 978-3-031-51893-5 (eBook)
https://doi.org/10.1007/978-3-031-51893-5

This Springer imprint is published by the registered company Springer Nature Switzerland AG
The registered company address is: Gewerbestrasse 11, 6330 Cham, Switzerland

Paper in this product is recyclable.

Contents

Understanding Concepts in Graph Signal Processing for Neurophysiological Signal Analysis

Stephan Goerttler, Min Wu, and Fei He

1 Introduction

Multivariate signals are composite signals that consist of multiple simultaneous signals, typically time signals. They can be either directly acquired using arrays of spatially separated sensors [1] or inferred using more complex recording techniques, such as magnetic resonance imaging [2]. The lowering cost of sensors has made multivariate signals ubiquitous, which highlights the importance to develop or improve tools to process these signals [3–5]. Examples of multivariate signals are as diverse as temperature measurements at different geographic locations [6], pixels in images or digital movies [7], or the measurement of process variables in nuclear reactors [8]. Multivariate signals are also common in various biomedical imaging applications, such as electroencephalography (EEG) [9], electrocorticography [10], magnetoencephalography (MEG) [11], or functional magnetic resonance imaging (fMRI) [12]. This book chapter will focus primarily on multivariate EEG signals.

Unlike time or spatial signals, multivariate signals can capture the characteristics of two dimensions of the measured system. In EEG, sensors are placed on various

---

S. Goerttler (✉)
Centre for Computational Science and Mathematical Modelling, Coventry University, Coventry, UK

Institute for Infocomm Research, Agency for Science, Technology and Research (A*STAR), Singapore
e-mail: goerttlers@uni.coventry.ac.uk

M. Wu
Institute for Infocomm Research, Agency for Science, Technology and Research (A*STAR), Singapore

F. He
Centre for Computational Science and Mathematical Modelling, Coventry University, Coventry, UK

A. Ahmed, J. Picone (eds.), *Machine Learning Applications in Medicine and Biology*,
https://doi.org/10.1007/978-3-031-51893-5_1

locations on the scalp of the brain [9]. Each sensor, also called *channel* in this context, measures the electric field potential in time at its location, giving information about the dynamics of the electric activity. On the other hand, the electric activity of all sensors at one point in time provides information about the spatial characteristics of the system.

EEG recordings are most commonly analysed as time signals using conventional signal processing [13]. Historically, one fruitful approach has been the use of *power spectral density* estimates to extract spectral features for each channel. These features can be used to quantify the power of widely recognised *brain waves*, such as alpha (8–12 Hz), beta (12–35 Hz), gamma (>35 Hz), delta (0.5–4 Hz), and theta (4–8 Hz) waves [14]. The extent of these brain waves, together with their position on the scalp, has been shown to be useful biomarkers for Alzheimer's disease [15], sleep apnea syndrome [16], social anxiety disorders [17], or even performance in sport [18]. However, numerous other features can be extracted from the time signals, which can be both linear and non-linear. For example, non-linear summary metrics such as the Lyapunov exponent may predict the onset of epileptic seizures [19]. A second, traditional method to analyse EEG recordings in time is to relate the electrical activity to specific events such as task stimuli [20]. The electric potential can be averaged to yield the so-called *event-related potential* (ERP). The amplitude of the ERP at specific time delays for relevant tasks has been shown to be affected by disorders such as alcoholism [21] or schizophrenia [22].

Alternatively, multivariate signals can be analysed along their spatial dimension. This type of analysis rests on the assumption that the measured system has a *connectivity structure*, either static or changing with time, which causes some of the signals to be pairwise correlated. For example, Peters et al. analysed the connectivity structures retrieved from EEG recordings with tools from *graph theory*, allowing them to characterise autism [23]. Similarly, Atasoy et al. analysed the connectivity structure in fMRI recordings using *spectral graph theory* and were able to link the resulting harmonic waves to connectome networks. A comprehensive review of graph analysis methods for neuroimaging applications is given by Fallani et al. [24].

However, both the temporal and the spatial analyses of multivariate signals neglect possible spatio-temporal features in these signals, i.d., features that are both present in the temporal and spatial domains. Due to this shortcoming, *graph signal processing* (GSP) has emerged as a new promising framework that analyses the signals in time in dependence on their graph structure [25, 26], thereby taking both domains into account simultaneously. One core idea behind GSP is to transform the data across space based on the connectivity structure, before processing the signals with conventional tools. This spatial transformation is also called the *graph Fourier transform* (GFT), following its classical counterpart. GSP has been successfully employed for *neurophysiological imaging*, such as fMRI [27–30] or EEG [31–33], which will be discussed in more detail in Sect. 3. Comprehensive review articles of GSP are those of Shuman et al. [34] and Ortega et al. [25], while Stanković et al. [5] may be a good resource for readers looking for a more intuitive introduction into GSP.

Currently, the ambiguity of the connectivity structure [35] and how this structure is used may be a significant limitation of GSP, given that the connectivity structure is an essential building block in this theory. Addressing this challenge will be a central theme in this book chapter. To this end, the theoretical section aims to derive GSP concepts from basic principles, juxtapose the alternative formulations in GSP, and explore links between them. On the other hand, the experimental section showcases a strong validation procedure using a novel baseline framework, which is needed in light of these ambiguities. It further shows how simulations can help to understand mechanisms in GSP by overcoming the data scarcity commonly observed in empirical data sets.

The experiment itself evaluates the role of *graph frequencies* of GFT-transformed signals, which may aid in GSP-based dimensionality reduction for neurophysiological signals. Our work complements a study by Ouahidi et al., which aimed to select the most valuable graph frequencies for fMRI decoding [36]. In our study, we classified each graph frequency signal separately using a *support vector machine* (SVM), allowing us to determine the classification accuracy with respect to graph frequency. Our analysis of the results, aided by the carefully designed baseline framework, finally revealed that the superior performance of high-frequency transformed signals can be attributed to the graph structure.

The book chapter is organised as follows: Sect. 2 gives a mathematical introduction into GSP by firstly covering graph terminology and basic principles, before deriving key concepts in GSP. Section 3 relates these theoretical concepts to neuroimaging. While the first three subsections cover the background for neuroimaging with GSP, the following subsections illustrate several applications of GSP in this field. Section 4 draws important links between the graph retrieval methods and graph representations, both generally and applied to neurophysiological signals. Section 5 commences the experimental part of this book chapter and covers the methodology of our experiment, comprising the simulation of the data, the classification model, the testing, and the baseline framework. The subsequent Sect. 6 presents the results of our study, while Sect. 7 discusses these results. The last Sect. 8 summarises the book chapter and proposes future directions.

2 Concepts in Graph Signal Processing

In this chapter, GSP concepts are developed in analogy to classical signal processing, following work by Sandryhaila et al. [37]. Figure 2 gives a comprehensive overview of the relations between the relevant concepts in classical signal processing and how they are extended to GSP. Importantly, this extension is ambiguous and leads to two separate definitions of the GFT, from which some of the other concepts are derived.

2.1 Graph Terminology

Formally, networks can be represented by a *weighted graph* $\mathcal{G} := (\mathcal{V}, \mathbf{A})$, which consists of N vertices $\mathcal{V} = \{1, 2, 3, \ldots, N\}$ and the *weighted adjacency matrix* $\mathbf{A} \in \mathbb{R}^{N \times N}$. The entries a_{ij} of $\mathbf{A}$ represent the strength of the connectivity between node i and node j. If the connectivities between each pair of nodes are symmetric, or $a_{ij} = a_{ji}$ for any i and j, then $\mathbf{A}$ is symmetric, and the graph is said to be undirected; in the other case, the graph is said to be directed. Furthermore, if the diagonal elements a_{ii} are all zero, meaning that the nodes are not connected to themselves with loops, the graph is called a *simple graph*. Figure 1a depicts a weighted directed simple graph with $N = 6$ vertices.

A second, useful representation of a graph $\mathcal{G}$ is its *Laplacian matrix*. The Laplacian can be directly computed from the adjacency matrix $\mathbf{A}$ as $\mathbf{L} = \mathbf{D} - \mathbf{A}$. Here, $\mathbf{D} = \text{diag}(\mathbf{A} \cdot \mathbf{1})$ denotes the *degree matrix*. Alternatively, the symmetric normalised Laplacian can be used instead of the Laplacian, which is computed as $\mathbf{L}_{\text{norm}} = \mathbf{D}^{-1/2}\mathbf{L}\mathbf{D}^{-1/2}$. While the weighted adjacency matrix can be viewed as a *graph shift operator* (GSO, Sect. 2.2), the Laplacian can be associated with the negative *difference operator* (Sect. 2.3).

A multivariate signal $\mathbf{X} \in \mathbb{R}^{N \times N_t}$ is a data matrix consisting of N univariate signals each of length N_t, typically temporal signals. We denote the univariate signal at node i by $\mathbf{x}_{i*}$. The rows of $\mathbf{X}$, on the other hand, correspond to *graph signals* and contain one value for each node on the graph. We denote the graph signals as either $\mathbf{x}$ or more explicitly as $\mathbf{x}_{*j}$, where j is the column index.

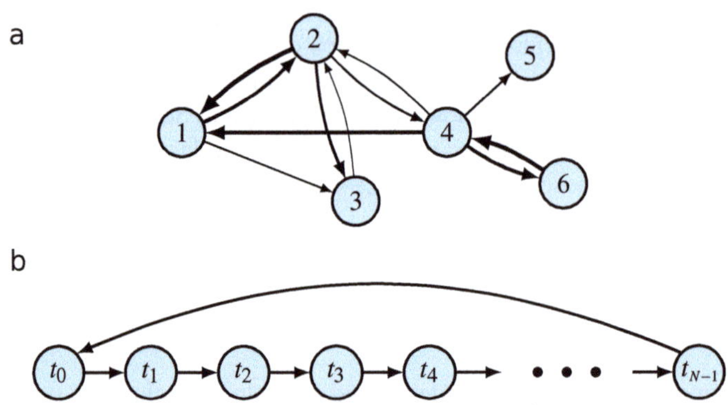

Fig. 1 (**a**) Graph with six nodes, which can be represented by an adjacency matrix **A**. The displayed connections are directed, meaning that they can go in both directions. They are also weighted, which is indicated by the line width of the arrows. The graph can represent an arbitrary spatial graph topology of a multivariate signal acquired from a measurement system with six sensors. (**b**) Time graph with linear topology. The graph connects each time step t_i to the subsequent time step t_{i+1}, thereby shifting a signal in time. For periodic signals, the last time step is connected to the first time step. The graph can be represented by the cyclic shift matrix $\mathbf{A}_c$

2.2 Graph Shift Operator (GSO)

The weighted adjacency matrix is the canonical algebraic representation of a graph. This subsection further explores the role of the adjacency matrix as a GSO and its implications for graph dynamics. In particular, these graph dynamics are capitalised for generating the artificial multivariate signals in Sect. 5.1.

In *discrete signal processing* (DSP), a cyclic time signal s with N samples can be represented algebraically by a time-ordered vector $\mathbf{s} = (s_0, \ldots, s_{N-1})^\top$, where s_0 and s_{N-1} are the time samples at the first and the last time step, respectively. The *shift operator* T_1 shifts each time sample s_n to the subsequent time step,

$$T_1 : (s_0, \ldots, s_{N-1})^\top \mapsto (s_{N-1}, s_0, \ldots, s_{N-2})^\top,$$

and can be algebraically represented by the cyclic shift matrix

$$\mathbf{A}_c = \begin{bmatrix} 0 & 0 & 0 & \ldots & 1 \\ 1 & 0 & 0 & \ldots & 0 \\ 0 & 1 & 0 & \ldots & 0 \\ \vdots & & \ddots & \ddots & \vdots \\ 0 & & \ldots & 1 & 0 \end{bmatrix}. \tag{1}$$

Note that the last time sample s_{N-1} is shifted to the first time step, as we assume the time signal to be cyclic.

Importantly, $\mathbf{A}_c$ can be interpreted as the adjacency matrix of a graph with a linear topology with N vertices, as shown in Fig. 1b. The interpretation of the discretised time dimension as a linear time graph allows to generalise the shift operator to more arbitrary graph topologies, such as the one shown in Fig. 1a. Consequently, the GSO of a graph is simply given by its weighted adjacency matrix $\mathbf{A}$. Applied to a graph signal $\mathbf{x}$, the adjacency matrix spatially shifts a signal at node i to its neighbouring nodes, weighted by the strength of the connection. The GSO may also be used to model graph dynamics. This follows the assumption that the time evolution from one time step to the next corresponds to a scaled graph shift, that is, the GSO is taken to be a "time evolution operator". However, this crude approximation only describes the dynamics of the signals which are due to the connectivity structure. Nevertheless, this characterisation of the GSO proves to be useful to simulate a multivariate signal in Sect. 5.1. A more sophisticated way to ascribe the graph dynamics to the connectivity structure is to model the dynamics as a heat diffusion process [35], which involves the Laplacian operator. The Laplacian operator itself is based on the GSO, as shown in the following Sect. 2.3.

2.3 Laplacian as Negative Difference Operator

In discrete calculus, the operator Δ defines a *forward difference* in terms of the shift operator T_1 and the identity operator I:

$$\Delta = T_1 - I.$$

In the case of a finite, cyclic signal, the shift operator can be represented algebraically by the cyclic shift matrix $\mathbf{A}_c$ as defined in Eq. (1), while the identity operator can be represented by the degree matrix $\mathbf{D}_c = \mathbb{1}$. The negative forward difference operator $-\Delta$ can then be represented by the Laplacian matrix $\mathbf{L}_c$:

$$\mathbf{L}_c = \mathbf{D}_c - \mathbf{A}_c.$$

Likewise, the second-order central difference operator Δ^2 is given in terms of the right shift operator T_{-1}:

$$\Delta^2 = T_1 - 2I + T_{-1}$$
$$T_{-1} : (s_0, \ldots, s_{N-1})^\top \mapsto (s_1, \ldots, s_{N-1}, s_0)^\top.$$

The right shift operator T_{-1} can be represented algebraically by the transposed adjacency matrix $\mathbf{A}_c^\top$, such that the operator $-2\Delta^2$ can be represented by the symmetrised Laplacian matrix

$$\mathbf{L}_{c,\text{sym}} = \mathbf{D}_c - \frac{\left(\mathbf{A}_c + \mathbf{A}_c^\top\right)}{2}.$$

2.4 Graph Fourier Transform (GFT)

The GFT is an extension of the Fourier transform to graphs and is at the heart of GSP. Here, we show that it can be derived either via graph filtering (*filter-based GFT*) [25, 38] or via discrete derivatives on graphs (*derivative-based GFT*) [38, 39]. Both these approaches are built on the extension of the shift operator to graphs, i.d., the GSO given by the adjacency matrix $\mathbf{A}$. While both approaches yield the same result in DSP, their extensions to GSP turn out to be slightly different: The derivative-based GFT is composed of the eigenvectors of the adjacency matrix, whereas the filter-based GFT is composed of the eigenvectors of the Laplacian matrix.

The two approaches mirror the two definitions for the total variation in Sect. 2.5. Section 4.2 shows theoretically why the two approaches can nevertheless be similar to each other in some practical applications.

Filter-Based Graph Fourier Transform

The derivation in this subsection follows Sandryhaila et al. [40] and Ortega et al. [25]. Let s be a time signal represented by a time-ordered vector $\mathbf{s} = (s_0, \ldots, s_{N-1})^\top$ and T_1 a shift operator represented algebraically by $\mathbf{A}_c$, as defined in Eq. (1). A *finite impulse response filter* h of order N can be characterised as a sum of the N samples $s_{in}[n-i \mod N]$ preceding the time step n, weighted by p_i:

$$s_{out}[n] = (h \cdot s_{in})[n] = \sum_{i=0}^{N-1} p_i s_{in}[n-i \mod N].$$

Importantly, this means that the filter h can be represented as a N-th order polynomial in the time shift operator T_1, as defined in Sect. 2.2:

$$h = \sum_{i=0}^{N-1} p_i T_1^i.$$

Algebraically, the filter is given by the matrix $\mathbf{H}$ as a polynomial in the cyclic shift matrix $\mathbf{A}_c$ as defined in Eq. (1):

$$\mathbf{H} = \sum_{i=0}^{N-1} p_i \mathbf{A}_c^i.$$

The eigenvectors of $\mathbf{A}_c$ are the complex exponentials $\mathbf{v}_k = (\omega^0, \omega^k, \ldots, \omega^{(N-1)k})^\top / \sqrt{N}$, with $\omega^k = e^{2\pi jk/N}$ and the imaginary unit $j = \sqrt{-1}$. The eigenvalue for each complex exponential $\mathbf{v}_k$ is ω^k. Consequently, the eigendecomposition of $\mathbf{A}_c$ is given by:

$$\mathbf{A}_c = \mathbf{Q}_c \Lambda_c \mathbf{Q}_c^{-1} = (\mathbf{v}_1, \mathbf{v}_2, \ldots, \mathbf{v}_N) \begin{bmatrix} \omega^0 & & & \\ & \omega^1 & & \\ & & \ddots & \\ & & & \omega^{N-1} \end{bmatrix} (\mathbf{v}_1^*, \mathbf{v}_2^*, \ldots, \mathbf{v}_N^*)^\top. \tag{2}$$

Crucially, the matrix $\mathbf{Q}_c^{-1}$, comprising the complex conjugated exponentials $\mathbf{v}_k$, can be identified as the discrete Fourier transform (DFT) matrix.

The eigenvectors $\mathbf{v}_k$ of $\mathbf{A}_c$ are also eigenvectors of the filter matrix $\mathbf{H}$,

$$\mathbf{H}\mathbf{v}_k = \sum_{i=0}^{N-1} p_i \mathbf{A}_c^i \mathbf{v}_k = \left(\sum_{i=0}^{N-1} p_i \left(\omega^k\right)^i \right) \mathbf{v}_k, \tag{3}$$

which of course is the well-known fact in DSP that filters can change the magnitude or the phase of a sinusoidal signal but not its frequency.

The central role that the shift operator plays in constructing the filter and its relation to the DFT justifies to define the GFT in terms of the GSO, which can be represented algebraically by the adjacency matrix $\mathbf{A}$ (see Sect. 2.2). In analogy to DSP, the filter-based GFT for a spatial signal $\mathbf{x} \in \mathbb{R}^N$ on a graph $\mathbf{A}$ is defined using the eigendecomposition $\mathbf{A} = \mathbf{Q_A} \Lambda_\mathbf{A} \mathbf{Q_A}^{-1}$:

$$\tilde{\mathbf{x}} := \mathbf{Q}_\mathbf{A}^{-1} \mathbf{x}, \tag{4}$$

where $\tilde{\mathbf{x}}$ is the transformed spatial signal. The eigenvalues λ_k of $\mathbf{A}$ are the diagonal elements of $\Lambda_\mathbf{A}$. If $\mathbf{A}$ is symmetric, they are real-valued and can be sorted in ascending order, and the eigenvectors $\mathbf{v}_k$ become orthogonal.

However, other matrices, such as the Laplacian matrix $\mathbf{L}_c = \mathbf{D}_c - \mathbf{A}_c$, have the same eigenvectors as $\mathbf{A}_c$ and can be similarly decomposed using the DFT matrix. Importantly, this degeneracy in the DSP case is lifted when the cyclic graph is extended to more complex graph topologies, as also illustrated in the overview in Fig. 2. This yields the alternative, derivative-based definition of the GFT, which is derived in the following subsection.

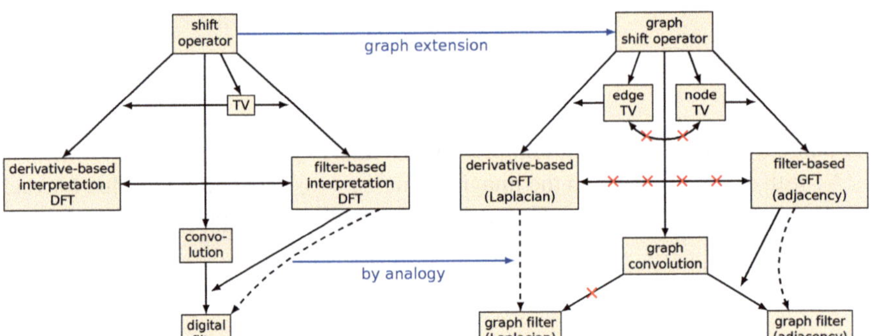

Fig. 2 Interconnections between discrete signal processing (DSP) concepts and its graph extensions. In DSP, the discrete Fourier transform (DFT) can be traced back to the shift operator. Crucially, a derivative-based interpretation and a filter-based interpretation of the DFT are equivalent. The total variation (TV) increases with increasing Fourier frequency, thereby linking the two concepts. Convolution and digital filters are built on the concept of the shift operator, whereby the digital filter can be expressed using the DFT. Extending the shift operator to graphs allows to build analogous concepts for graphs. Importantly, the derivative- and the filter-based graph Fourier transforms (GFTs) are not equivalent. The graph Fourier modes in both GFT variants can be ordered by their frequency using either the edge-based or the node-based TV. The graph convolution can be used to define an adjacency matrix-based graph filter. The similar, Laplacian matrix-based graph filter can only be constructed by analogy and is not directly linked to the graph convolution

Derivative-Based Graph Fourier Transform

In the classical Fourier transform, *Fourier modes* are complex exponentials $e^{-j\omega t}$. These exponentials are trivially eigenfunctions of the partial time-derivative $\partial/\partial t$ [38, 39], as well as the second partial time-derivative $\partial^2/\partial t^2$:

$$\frac{\partial}{\partial t}e^{-j\omega t} = -j\omega e^{-j\omega t}$$
$$\frac{\partial^2}{\partial t^2}e^{-j\omega t} = -\omega^2 e^{-j\omega t},$$

with eigenvalues of $-j\omega$ and $-\omega^2$, respectively.

As shown in Sect. 2.3, in DSP the directed Laplacian matrix $\mathbf{L}_c$ of a cyclic shift graph is the algebraic representation of the negated forward difference operator $-\Delta$, whereas the symmetrised Laplacian matrix $\mathbf{L}_{c,\text{sym}}$ is the algebraic representation of the operator $-2\Delta^2$. Crucially, the eigenvectors $\mathbf{v}_k$ of $\mathbf{A}_c$, which constitute the discrete Fourier modes, are also eigenvectors of $\mathbf{L}_c$ with eigenvalues $\lambda'_k = (1 - \lambda_k)$:

$$\mathbf{L}_c\mathbf{v}_k = (\mathbf{D}_c - \mathbf{A}_c)\,\mathbf{v}_k = 1 \cdot \mathbf{v}_k - \lambda_k\mathbf{v}_k = \lambda'_k\mathbf{v}_k.$$

In other words, the discrete Fourier modes are eigenvectors of the negated forward difference operator, in the same way that continuous Fourier modes are eigenfunctions of the time derivative. As a consequence, the eigendecomposition of $\mathbf{L}_c$ is given by:

$$\mathbf{L}_c = \mathbf{Q}_c^{-1}(\mathbb{1} - \Lambda_c)\,\mathbf{Q}_c =: \mathbf{Q}_c^{-1}\Lambda'_c\mathbf{Q}_c,$$

where Λ_c and $\mathbf{Q}_c$ are the eigenvalue and the eigenvector matrix of $\mathbf{A}_c$, respectively, as defined by Eq. (2).

The Fourier modes can be interpreted as eigenmodes of the derivative, or in the discrete case, as eigenmodes of the forward difference. Leveraging this notion leads to the second extension of DFT to graphs. Accordingly, the derivative-based GFT for a spatial signal $\mathbf{x} \in \mathbb{R}^{N_c}$ on a graph with adjacency matrix $\mathbf{A}$ is defined as follows:

$$\mathbf{L} = \mathbf{D} - \mathbf{A} = \mathbf{Q_L}\Lambda_\mathbf{L}\mathbf{Q_L}^{-1}$$
$$\tilde{\mathbf{x}} := \mathbf{Q_L}^{-1}\mathbf{x}. \tag{5}$$

In the literature, the symmetrically normalised Laplacian $\mathbf{L}_\text{norm}$ is sometimes used instead of the Laplacian [25, 38, 41]. In the case of a graph $\mathbf{A}_c$ with linear topology, the degree matrix $\mathbf{D}_c$ is given by the identity matrix, which means that the symmetrically normalised Laplacian $\mathbf{L}_{c,\text{norm}} = \mathbf{D}_c^{-1/2}\mathbf{L}_c\mathbf{D}_c^{-1/2}$ of this graph is equivalent to the Laplacian $\mathbf{L}_c$. Consequently, the eigendecomposition of $\mathbf{L}_{c,\text{norm}}$ is also given in terms of the DFT matrix.

Lastly, note that eigenvectors $\mathbf{v}_k$ of $\mathbf{L}_c$ with $k > 0$ are not the same as the eigenvectors of the symmetrised Laplacian $\mathbf{L}_{c,\text{sym}} = \mathbf{D}_c - \left(\mathbf{A}_c + \mathbf{A}_c^\top\right)/2$: While the eigenvectors $\mathbf{v}_k$ are complex-valued for $k > 0$, the eigenvectors of the $\mathbf{L}_{c,\text{sym}}$ are necessarily real-valued due to the symmetry of $\mathbf{L}_{c,\text{sym}}$. As a consequence, the eigendecomposition of $\mathbf{L}_{c,\text{sym}}$ is not given in terms of the DFT matrix; in other words, the eigendecomposition of the symmetrised Laplacian of a graph with linear topology does not reduce to the DFT. However, given the analogy of $\mathbf{L}_{c,\text{sym}}$ to the second partial time derivative in the continuous case, of which the Fourier modes are also eigenfunctions, it may still be valid to use the symmetrised Laplacian $\mathbf{L}_{\text{sym}}$ of an arbitrary graph structure for the GFT in lieu of $\mathbf{L}$. This may be especially useful if the measurement of the graph structure is intrinsically symmetric, while the actual graph structure is not. For example, when assessing the structural connectivity by measuring the white matter between brain regions, the directionality cannot be determined due to limitations in the methodology, and only the symmetric graph is retrieved.

In both variants, the GFT is a linear spatial transformation that is retrieved by computing the eigendecomposition of an algebraic graph representation. The rows of the GFT-matrix, which are the eigenvectors of the eigendecomposition, represent the *graph Fourier modes*. The computational complexity of this eigendecomposition is small as long as the number of graph nodes is sufficiently small, which is typically the case for neurophysiological applications. The GFT itself corresponds to a matrix multiplication with matrices of size $N \times N$ and $N \times N_t$, thereby scaling linearly with the number of time samples N_t.

While Eqs. (4) and (5) define the GFT for spatial signals, the GFT can be straightforwardly extended to multivariate signals $\mathbf{X} \in \mathbb{R}^{N \times N_t}$:

$$\tilde{\mathbf{X}} = \mathbf{Q}_{\mathbf{A}/\mathbf{L}}^{-1}\mathbf{X},$$

where the subscript $\mathbf{A}/\mathbf{L}$ indicates that the GFT can either be chosen to be filter-based ($\mathbf{A}$, computed using the adjacency matrix) or derivative-based ($\mathbf{L}$, computed using the Laplacian matrix). In the matrix multiplication on the right-hand side, each column $\mathbf{x}_{*j}$, corresponding to a spatial signal at time j, is transformed successively. The rows $\tilde{\mathbf{x}}_{i*}$ of the transformed multivariate signal $\tilde{\mathbf{X}}$ are the *graph frequency signals*, associated with the graph frequency i and the eigenvalue λ_i. Note that each graph frequency signal can also be thought of as a linear combination of the N time signals $\mathbf{x}_{i*}$.

2.5 Total Variation (TV)

A useful statistic of a graph signal $\mathbf{x} \in \mathbb{R}^{N_c}$ is its *total variation* (TV), which is a measure of how smooth the signal is across the graph structure $\mathbf{A}$. The TV is based on the local variation, which in turn is based on the graph derivative of the graph signal. The graph derivative can be defined either on the node i or on the edge (i, j), leading to two different definitions of the TV, here referred to as the *node-based* [38] and the *edge-based* [34] TV.

The node derivative calculates the difference between the signal and the signal shifted by the GSO, yielding a derivative at node i:

$$\nabla_i(\mathbf{x}) := [\mathbf{x} - \mathbf{A}\mathbf{x}]_i = x_i - \sum_j a_{ij} x_j.$$

The local variation is then given by the magnitude $\|\nabla_i(\mathbf{x})\|_1$ of the graph derivative.

The edge derivative, on the other hand, weighs the difference between the signal at node i and the signal at node j by their connectivity, yielding a derivative along the edge (i, j):

$$\nabla_{ij}(\mathbf{x}) := \sqrt{a_{ij}}(x_i - x_j).$$

The local variation is then computed using the p-norm of the derivative vector $\nabla_{i*}(\mathbf{x})$:

$$\|\nabla_{i*}(\mathbf{x})\|_p = \left(\sum_j a_{ij}^{\frac{p}{2}} |x_i - x_j|^p\right)^{\frac{1}{p}}, \quad p \in \mathbb{N}.$$

In both cases, the TV of a graph signal is computed from the sum of the local variation across all nodes. Given a graph structure $\mathbf{A}$, the node-based [38] and the edge-based [34] TV of a graph signal $\mathbf{x}$ are then defined as follows:

$$\begin{aligned}
\mathrm{TV}_{\mathbf{A}}^{(n)}(\mathbf{x}) &:= \sum_i \|\nabla_i(\mathbf{x})\|_1 = \sum_i \left| x_i - \sum_j a_{ij} x_j \right| = \|\mathbf{x} - \mathbf{A}\mathbf{x}\|_1 \\
\mathrm{TV}_{\mathbf{A}}^{(e)}(\mathbf{x}) &:= \frac{1}{2} \sum_i \|\nabla_{i*}(\mathbf{x})\|_2^2 = \frac{1}{2} \sum_i \sum_j a_{ij}(x_i - x_j)^2.
\end{aligned} \tag{6}$$

Note that the node-based TV can also be computed from the normalised adjacency matrix $\mathbf{A}_{\mathrm{norm}} = \mathbf{A}/|\lambda_{\max}|$, where $\lambda_{\max}$ is the largest eigenvalue of $\mathbf{A}$, which ensures numerical stability. While the node-based TV is always either positive or zero, the edge-based TV can become negative if we allow negative weights $a_{ij} < 0$ in the adjacency matrix. Section 3.1 discusses the validity of negative weights and possible implications.

The TV allows to understand the ordering of the graph eigenvectors as frequencies in terms of their eigenvalue. In the case of the node-based TV with $\mathbf{A}_{\mathrm{norm}}$, the TV for a normalised eigenvector $\mathbf{v}_k$ with real eigenvalue λ_k is given by:

$$\mathrm{TV}_{\mathbf{A}_{\mathrm{norm}}}^{(n)}(\mathbf{v}_k) = \left| 1 - \frac{\lambda_k}{\lambda_{\max}} \right|.$$

It can easily be seen that the following holds for two eigenvectors $\mathbf{v}_k$ and $\mathbf{v}_l$ of $\mathbf{A}_{\text{norm}}$ with respective real eigenvalues λ_k and λ_l:

$$\begin{aligned} &\lambda_k < \lambda_l \\ \Rightarrow\ &\mathrm{TV}^{(n)}_{\mathbf{A}_{\text{norm}}}(\mathbf{v}_k) > \mathrm{TV}^{(n)}_{\mathbf{A}_{\text{norm}}}(\mathbf{v}_l). \end{aligned}$$

In other words, the ordering of the eigenvalues from highest to lowest orders the graph eigenvectors from lowest to highest frequency. Notice that any eigenvector $\mathbf{v}_k$ of $\mathbf{A}_{\text{norm}}$ is also an eigenvector of $\mathbf{A}$.

A similar relationship between eigenvalue and frequency can be established for the edge-based TV if we assume the adjacency matrix to be symmetric, i.d., $a_{ij} = a_{ji}$. To this end, the TV is firstly expressed in terms of the Laplacian:

$$\begin{aligned} \mathrm{TV}^{(e)}_{\mathbf{A}}(\mathbf{x}) &= \frac{1}{2}\sum_{i,j} a_{ij}(x_i - x_j)^2 \\ &\overset{a_{ij}=a_{ji}}{=} 2\left(\frac{1}{2}\sum_{i,j} a_{ij}x_i^2\right) - \frac{1}{2}\sum_{i,j} a_{ij}2x_ix_j \\ &= \mathbf{x}^\top \begin{bmatrix} \sum_j a_{1j} & & \\ & \ddots & \\ & & \sum_j a_{N_cj} \end{bmatrix} \mathbf{x} - \mathbf{x}^\top \mathbf{A}\,\mathbf{x} = \mathbf{x}^\top \mathbf{L}\,\mathbf{x}. \end{aligned}$$

This representation of the TV allows to link the eigenvalues to graph frequencies. For a normalised eigenvector $\mathbf{v}_k$ of $\mathbf{L}$ with eigenvalue λ_k, the TV is given by:

$$\mathrm{TV}^{(e)}_{\mathbf{A}}(\mathbf{v}_k) = \mathbf{v}_k^\top \mathbf{L}\,\mathbf{v}_k = \lambda_k \|\mathbf{v}_k\|_2 = \lambda_k.$$

Hence, the frequency of the eigenvector $\mathbf{v}_k$ of $\mathbf{L}$, which is given in terms of its TV, is directly linked to its eigenvalue. Importantly, here the eigenvectors of the Laplacian matrix are linked to the graph frequencies, while in the case of the node-based TV the ones of the adjacency matrix are linked to the graph frequencies. This explains why here higher eigenvalues correspond to higher graph frequencies, while in the case of the node-based TV, the relationship is inversed.

If the Laplacian matrix is symmetric, the eigendecomposition is given by $\mathbf{L} = \mathbf{Q}_{\mathbf{L}} \Lambda_{\mathbf{L}} \mathbf{Q}_{\mathbf{L}}^\top$, and the expression can be further rewritten as follows:

$$\begin{aligned} \mathrm{TV}^{(e)}_{\mathbf{A}}(\mathbf{x}) &= \mathbf{x}^\top \mathbf{L}\,\mathbf{x} = \mathbf{x}^\top \mathbf{Q}_{\mathbf{L}} \Lambda_{\mathbf{L}} \mathbf{Q}_{\mathbf{L}}^\top \mathbf{x} \\ &= \tilde{\mathbf{x}}^\top \Lambda_{\mathbf{L}}\, \tilde{\mathbf{x}} = \sum_k \lambda_k \tilde{x}_k^2. \end{aligned}$$

This identity of the edge-based TV allows to understand how the TV of a graph signal can be decomposed in terms of its graph frequency components $\tilde{x}_k$.

2.6 Graph Convolution

The definition of the *graph convolution* is built on the graph shift operator defined in Sect. 2.2. In the classical case, the convolution for two period time signals s and r, represented by the time-ordered vectors $\mathbf{s} = (s_0, \ldots, s_{N-1})^\top$ and $\mathbf{r} = (r_0, \ldots, r_{N-1})^\top$, is given by the vector

$$\Big((s * r)[n]\Big)^\top_{1 \le n \le N} = \left(\sum_i r_i \mathbf{A}_c^i\right) \mathbf{s}.$$

The following two identities can be directly derived from the eigendecomposition of $\mathbf{A}_c$ in Eq. (2):

$$\begin{aligned} \mathbf{A}_c^i &= \mathbf{Q}_c \Lambda_c^i \mathbf{Q}_c^{-1} \\ \sum_i r_i \Lambda_c^i &= \operatorname{diag}(\mathbf{Q}_c^{-1}\mathbf{r}/\sqrt{N}). \end{aligned} \tag{7}$$

Using these two identities, the vector of the convoluted signal $s * r$ can be rewritten as follows:

$$\begin{aligned} \Big((s * r)[n]\Big)^\top_{1 \le n \le N} &= \sum_i r_i \, \mathbf{Q}_c \Lambda_c^i \mathbf{Q}_c^{-1} \mathbf{s} \\ &= \mathbf{Q}_c \operatorname{diag}\left(\mathbf{Q}_c^{-1}\mathbf{r}/\sqrt{N}\right) \mathbf{Q}_c^{-1}\mathbf{s} \\ &= \mathbf{Q}_c \frac{1}{\sqrt{N}} (\mathbf{Q}_c^{-1}\mathbf{r}) \circ (\mathbf{Q}_c^{-1}\mathbf{s}). \end{aligned}$$

The analogy between the classical shift operator and the GSO can be used to define the graph convolution on two spatial signals $\mathbf{x}$ and $\mathbf{y}$:

$$\mathbf{x} * \mathbf{y} = \left(\sum_i y_i \mathbf{A}^i\right) \mathbf{x}. \tag{8}$$

Using the filter-based GFT, we can use the identity

$$\mathbf{A}^i = \mathbf{Q}_\mathbf{A} \Lambda_\mathbf{A}^i \mathbf{Q}_\mathbf{A}^{-1}, \tag{9}$$

which can be trivially derived from the eigendecomposition of **A**, to rewrite the graph convolution:

$$\mathbf{x} * \mathbf{y} = \sum_i y_i \, \mathbf{Q_A} \Lambda_{\mathbf{A}}^i \mathbf{Q_A}^{-1} \mathbf{x}$$

$$= \mathbf{Q_A} \operatorname{diag}\left(\sum_i y_i \lambda_0^i, \ldots, \sum_i y_i \lambda_{N-1}^i \right) \mathbf{Q_A}^{-1} \mathbf{x}.$$

The crucial difference between the classical convolution and the graph convolution is that the identity (7) does not have a comparable equivalent in GSP.

We can also implicitly define the graph convolution for the case of the derivative-based GFT by analogy to the classical convolution:

$$\mathbf{x} * \mathbf{y} = \mathbf{Q_L} \left(\sum_i y_i \Lambda_{\mathbf{L}}^i \right) \mathbf{Q_L}^{-1} \mathbf{x} = \left(\sum_i y_i \mathbf{L}^i \right) \mathbf{x}. \tag{10}$$

This definition of the graph convolution has been commonly used in the literature [42–45]. Note, however, that this definition is implicitly based on the inadequate assumption that the Laplacian matrix is a GSO, which can be seen by comparing the last Eq. (10) to the GSO-based convolution in Eq. (8).

2.7 Graph Signal Filtering

In classical signal processing, filters can be described both by their impulse response in the time domain and by their frequency response in the Fourier frequency domain; the two descriptions are linked by the DFT. *Graph filters* can be similarly defined by their impulse and by their frequency response; here the descriptions are linked by the GFT [46]. The former, impulse response description was previously encountered in Sect. 2.4 and is closely related to the concept of convolution as introduced in the previous Sect. 2.6. The frequency response description is more convenient to define graph spectral band-pass filters, which is essential for tasks such as graph denoising. For example, a low-pass filter can be used to reduce the total variation of the spatial signal. Alternatively, graph impulse response filters can also be used to minimise the node-based TV together with a similarity term, as shown in Chen et al. [47]. However, this approach is less intuitive, as it requires to explicitly compute the filter parameters.

In the time domain, a filter **H** can be built by accessing previous signal values through shifting the signal **x** and weighing each of these values by p_i, as previously shown in Eq. (3):

$$\mathbf{H}\mathbf{x} = \left(\sum_{i=0}^{N-1} p_i \mathbf{A}_c^i \right) \mathbf{x} = h(\mathbf{A}_c)\mathbf{x} = \tilde{\mathbf{x}}, \tag{11}$$

where $\tilde{\mathbf{x}}$ is the filtered graph signal, and h is the polynomial filter function.

Using the identity (9) with $\mathbf{A} = \mathbf{A}_c$,

$$\mathbf{A}_c^i = \mathbf{Q}_c \Lambda_c^i \mathbf{Q}_c^{-1},$$

we can rewrite Eq. (11):

$$\begin{aligned} \mathbf{H}\mathbf{x} &= \mathbf{Q}_c \left(\sum_{i=0}^{N-1} p_i \Lambda_c^i \right) \mathbf{Q}_c^{-1}\mathbf{x} = \mathbf{Q}_c \, h(\Lambda_c) \, \mathbf{Q}_c^{-1}\mathbf{x} \\ &=: \mathbf{Q}_c \operatorname{diag}(h_0, \ldots, h_{N-1}) \, \mathbf{Q}_c^{-1}\mathbf{x}. \end{aligned}$$

In other words, to define the filter $\mathbf{H}$ we can either use the N parameters p_i, which define the filter by its weights in the time domain, or equivalenty the parameters h_i, which define it by its weights in the Fourier domain.

In analogy to this classical case, we can define a graph filter $\mathbf{H}_{\mathbf{A}/\mathbf{L}}$ in the spatial domain for the two definitions of the GFT as follows:

$$\begin{aligned} \mathbf{H}_{\mathbf{A}}\mathbf{x} &= \left(\sum_{i=0}^{N-1} p_i \mathbf{A}^i \right) \mathbf{x} \\ \mathbf{H}_{\mathbf{L}}\mathbf{x} &= \left(\sum_{i=0}^{N-1} p_i \mathbf{L}^i \right) \mathbf{x}. \end{aligned}$$

Note that the Laplacian-based GFT filter is not built on the graph convolution, unlike the adjacency matrix-based GFT filter.

As in the classical case, the same filters can be described in the graph spectral domain using the eigenvalue matrix $\Lambda_{\mathbf{A}/\mathbf{L}}$:

$$\begin{aligned} \mathbf{H}_{\mathbf{A}/\mathbf{L}}\mathbf{x} &= \mathbf{Q}_{\mathbf{A}/\mathbf{L}} \, h(\Lambda_{\mathbf{A}/\mathbf{L}}) \, \mathbf{Q}_{\mathbf{A}/\mathbf{L}}^{-1}\mathbf{x} \\ &=: \mathbf{Q}_{\mathbf{A}/\mathbf{L}} \operatorname{diag}\left(h_0^{(A/L)}, \ldots, h_{N-1}^{(A/L)} \right) \mathbf{Q}_{\mathbf{A}/\mathbf{L}}^{-1}\mathbf{x}. \end{aligned}$$

The parameters p_i or h_i can either be designed or learned using machine learning approaches. Sometimes, the spatial domain description of the filter is used to avoid computing the eigendecomposition of the adjacency matrix or the Laplacian [42]. To further limit computations or the number of parameters, only the first k parameters p_i are used, such that $p_i = 0$ for $i >= k$.

3 Graph Signal Processing in Biomedical Imaging

Biomedical imaging is one of many applications where multivariate signals are acquired. Brain imaging techniques, such as fMRI, EEG, or MEG, record multiple signals simultaneously in time at different spatial locations. In fMRI, the signal locations are called *voxels*, whereas in MEG and EEG, they are called *channels*. An EEG measurement setup with N_c channels which record the brain over a time period T at a sampling rate f yields a data matrix $\mathbf{X} \in \mathbb{R}^{N_c \times N_t}$, where $N_t = \lfloor f \cdot T \rfloor$ is the number of time samples. The following three sections cover graphs and graph Fourier modes in biomedical imaging more generally, before discussing applications of GSP in this field.

3.1 Graph Retrieval

The GFT of the multivariate signals relies on the graph, or spatial structure, of the data. The retrieval of the graph from neuroimaging data is not unique, but can be based on the structural, or anatomical, connectivity, the functional connectivity [48], or the geometric location of the sensors [27].

Firstly, the graph can be based on the *functional connectivity* between the signals. This method is purely data-driven and can therefore be used on all data sets. Common choices to build the graph include computing pairwise Pearson correlations or covariances, but nonlinear measures such as mutual information, the phase lage index, or the phase locking value can be used as well [49]. As shown in Sect. 4.3, there is a link between GFT using functionally retrieved graphs and PCA. The experimental section of this chapter uses the Pearson correlation to construct the graph (see Sect. 5.2). Secondly, a graph can be constructed using the *structural connectivity* between nodes, which can be determined with secondary measurements. For neurophysiological signals, for example, diffusion tensor imaging can be employed [50]. Lastly, the connectivity between the nodes can be determined by their *geometric* properties, such as the pairwise distances between them. The distances can then be mapped to the connectivity strengths. The method only requires knowledge about the data acquisition system.

Often, these different graph retrieval methods can lead to similar graphs, in part because the methods are aiming to estimate the same connectivity structure underlying the neural substrate. As an example of this, the structural and the functional connectivities are generally related to each other [50]. In an attempt to make this relation more explicit, Li et al. learned the mapping between structural and functional connectivities, giving insight into how functional connectivity arises from structural connectivity [51]. On a more cautious note, Wang et al. give examples why the structural connectivity is not sufficient to explain the dynamics of neural activity [52] and thus the functional connectivity.

Despite some similarities, Horwitz argued that there is no single underlying connectivity structure, but that connectivity should be thought of as "forming a class of concepts with multiple members" [53]. Generally, the choice of the graph retrieval method can alter the interpretation of the graph [54]. For example, Mortaheb et al. used a geometric graph to calculate the edge-based total variation, which allowed them to interpret the results as effects of local communication [31]. Nevertheless, the ambiguity of the retrieved graph remains a challenge for GSP. To illustrate this, Ménoret et al. observed vastly different results when testing seven different graphs for their GSP-based methodology [27]. An interesting solution to reduce this ambiguity is the construction of a multilayer graph, as was done by Cattai et al. with EEG data [55].

One important difference between the graph retrieval methods lies in whether they allow negative connections between nodes or not. For example, some functional connectivity measures, such as the pairwise Pearson correlation, can yield negative connections. On the other hand, structural connectivity measures looking at physical structures in the brain, such as white matter projections, typically cannot discern whether a connection is positive or negative [56]. The implications of using negative weights in the adjacency matrix are further explored in Sect. 3.2.

3.2 Negative Graph Weights

Some graph retrieval methods, such as the pairwise Pearson correlation, can lead to negative weights in the adjacency matrix. While negative weights may be less intuitive, they can extend the applicability of graphs [57]. In neuroimaging, for example, negative weights may be used to model inhibitory pathways in the brain [58]. This section will discuss some of the mathematical implications of negative weights for GSP.

A first consequence of using a graph with negative weights is that the Laplacian matrix is not necessarily diagonally dominant, which in turn means that the Laplacian can have negative eigenvalues. A downside of this is that the constant, or DC, graph Fourier mode is no longer the lowest Fourier mode, i.d., the mode with the lowest frequency. An example of this can be seen in Fig. 3, where the graph Fourier modes for two graphs are given, namely the geometric graph with only positive weights and the Pearson correlation graph with both positive and negative weights. The DC graph Fourier mode is the lowest mode for the geometric graph, which is not the case for the Pearson correlation graph. Secondly, allowing negative weights means that the edge-based TV, as defined in Eq. (6), can become negative. This is in contrast with the notion of TV as a positive quantity in the field of image processing [59]. A third consequence of negative weights is that they can cause negative entries in the degree matrix $\mathbf{D}$, which means that the square root of the inverse of $\mathbf{D}$, $\mathbf{D}^{-1/2}$, is not real-valued. As a result of this, the symmetric normalised Laplacian $\mathbf{L}_{\text{norm}} = \mathbf{D}^{-1/2}\mathbf{L}\mathbf{D}^{-1/2}$, which may be required for certain applications, can not be computed.

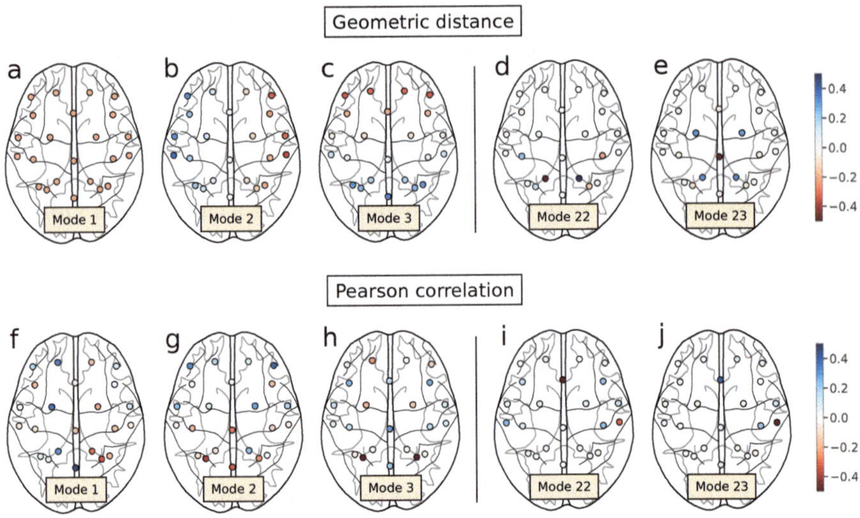

Fig. 3 Lowest and highest graph Fourier modes for a geometric distance-based graph and a functional connectivity-based graph, computed from a real EEG data set and visualised using the Nilearn package in Python [60]. (**a–c**) The lowest graph Fourier modes of the geometric graph capture the fundamental symmetries and comprise the DC mode, the lateral symmetry mode, and the coronal symmetry mode. (**d**, **c**) The highest graph Fourier modes are more localised. (**f–h**) The lowest graph Fourier modes of a functional connectivity Pearson correlation graph, which is computed from a real-world EEG data set. The graph contains negative weights, such that the DC mode is not necessarily the lowest mode. Modes 1 and 2 vaguely reflect the lateral symmetry mode 2 and the coronal symmetry mode 3 of the geometric distance-based graph. (**i**, **j**) The highest graph Fourier modes of the Pearson correlation graph are localised and still exhibit a lateral symmetry

3.3 Graph Fourier Modes

In *spectral graph theory*, the graph is decomposed into graph Fourier modes. The GFT transforms a spatial signal $\mathbf{x} \in \mathbb{R}^{N_c}$ by projecting it onto the N_c graph Fourier modes, which can be thought of as eigenmodes of the graph. This section intends to shed more light on the meaning of these graph Fourier modes.

Figure 3 shows the graph Fourier modes for a real EEG data set. The data set was recorded on 20 healthy participants and 20 Alzheimer's patients. The recordings were acquired using 23 bipolar channels at a sampling rate of 2048 Hz. For all participants, three 12 s-long segments were selected from a larger recording. For one participant, one segment is missing. A more detailed description of the data set can be found in Blackburn et al. [61]. The top row (a–e) in the figure shows graph Fourier modes computed from the geometric distance between the EEG sensors. The bottom row (f–j) shows modes computed from the functional connectivity, specifically the pairwise Pearson correlation between the sensors, averaged across all participants and segments. Here, the relation between the graph Fourier modes and their corresponding frequency becomes clearer: Lower frequency modes are "waves" with a global pattern spreading across the whole brain, whereas higher modes are highly localised and fast-varying waves. This general description applies to the two graphs, even though they are retrieved with fundamentally different methods. One major difference between the two graphs is that the DC mode, i.d., the mode with constant entries, is the lowest frequency mode for the geometric distance graph, but not for the Pearson correlation graph. This is a consequence of the negative edge weights in the correlation-based graph. These induce eigenmodes with negative eigenvalues and therefore have a frequency below that of the DC mode with eigenvalue zero.

The projections of a multivariate signal onto these graph Fourier modes, namely the *graph frequency signals*, may yield insight into the acquired data beyond the raw signal. For example, the projection of the multivariate signal onto the DC mode corresponds to the mean of the signal across the channels. The analysis of the graph frequency signals in terms of their performance on a classification task may shed more light on the meaning of graph frequency and is the main topic in the experimental part of this book chapter (see Sects. 5–7).

One major difference between DFT Fourier modes and GFT graph Fourier modes is the magnitude of the eigenvalues associated with these modes, which can take on values other than 1 in the case of GFT modes. Therefore, when using the GSO as a time evolution operator, the eigenmode coefficients may either vanish or explode. This is in contrast to the shift operator in Euclidean space, which ensures that eigenmode coefficients stay constant in time. One example from physics is electromagnetic waves, which can travel distances of billions of light years due to this property. Previously, an energy-preserving GSO was introduced by Gavili et al. [41]; however, this GSO is constructed by simply replacing the eigenvalues of the decomposed GSO with 1. While this changes the original GSO, it leaves the GFT and thereby the eigenmodes untouched. The new GSO preserves the energy of the eigenmodes, but it may no longer be an accurate representation of the graph structure.

3.4 Graph-Based Dimensionality Reduction

The GFT projects a spatial or a multivariate signal onto graph Fourier modes, which were the topic of the previous Sect. 3.3. These projections are called the graph frequency signals. Unlike the unaltered time signals, the graph frequency signals can be ordered. This ordering allows to reduce the dimensionality of the signal by selecting specific graph frequencies, such as the k lowest graph frequencies, as for example, demonstrated by Rui et al. on an MEG data set [62]. The same dimensionality reduction method can also be used for *graph-based image compression*, as shown by Fracastoro et al. [63]. The experimental section of this book chapter will further examine the role that the frequency ordering plays for the graph frequency signals.

Similar as in classical signal processing, graph bandlimited signals correspond to subsampling in the spatial domain [64]. However, unlike in the classical case, loss-free graph sampling for a given band requires sophisticated algorithms. Examples are the greedy algorithm by Anis et al. [64] and the random sampling scheme by Puy et al. [65]. Both the projection-based and the graph sampling-based dimensionality reduction were evaluated by Ménoret et al. on an fMRI classification task [27]. Their results indicate that improvements over graph-free dimensionality reduction methods depend on the graph used. We show in Sect. 4.3 that GSP-based dimensionality reduction is equivalent to *principal component analysis* (PCA) for a certain class of graphs.

3.5 Graph Filtering

Graph filtering, formally derived in Sect. 2.7, is a transformation that can be used for tasks such as graph denoising of a multivariate signal [47]. Graph denoising can be based on the graph frequency response filter using either the adjacency or the Laplacian matrix. In analogy to band-pass Fourier filtering [66], a low-pass graph filter can be applied to the input signal for graph denoising. Intuitively, structurally connected sensors measure similar values; therefore, variations between highly connected sensors should indicate noise. More formally, it has been shown that graph filters minimise the TV [34]. Conversely, variations between connected sensors may carry important information as well, in which case a high-pass graph filter would be more suitable. The graph-filtered signal can then be used for further signal processing tasks. Huang et al. use both a graph low-pass and a graph high-pass filter on fMRI signals, and find that the high-pass filter is more useful to correlate signal norms with switch cost in the Navon switching task. Graph band-pass filtering is also the basis for the structural-decoupling index introduced by Preti et al., which compares the norm of graph high-pass and low-pass filtered graph signals [67].

Graph impulse response filters can also be used for graph denoising, albeit less inuitively so. Chen et al. applied graph denoising on temperature sensor measurements and for opinion combining. A more sophisticated approach is to adaptively learn the graph filter, which has been used to denoise impulsive EEG signals [68].

Use in Neural Networks

In machine learning applications, graph filtering allows to extend convolutional neural networks to data on graphs. Introduced by Defferrard et al. [42], graph neural networks have found widespread applications [69]. Here, a Laplacian-based graph impulse response filter as opposed to a graph frequency response filter is commonly used, which has the advantage that it can be localised on the graph. To decrease the range of suitable filter values, the polynomial of the filter is commonly replaced by a Chebyshev polynomial. The polynomial impulse filter can also be replaced by an auto-regressive moving average filter, as shown by Bianchi et al. [70]. In EEG, graph neural networks have been employed to detect emotions [71, 72], to classify error-related potentials [73], or to detect seizures [74].

3.6 Analysis of Total Variation (TV)

The TV, as introduced in Sect. 2.5, is a statistic which measures how much a signal varies on the graph structure. While the previously examined spatial low-pass filter aims to reduce the TV of a signal on a graph, the TV in itself serves as a useful statistic of the graph signal or, when summed over time, the whole multivariate signal. For example, Mortaheb et al. found a correlation between the level of consciousness and the TV [31]. Specifically, the authors analysed alpha-band signals in EEGs on a geometric graph derived from the location of the sensors, giving insight into how disorders of consciousness affect short-range and long-range communication in the brain. The TV can also be used to extract features for classification tasks, such as motor imagery tasks from EEG [75].

3.7 Graph Learning with Graph Signal Processing (GSP)

Similar to graph denoising, GSP-based *graph learning* exploits the link between the graph structure and the spatial variation of a signal, i.d., the notion that a signal does not vary much on the graph structure. However, instead of changing the signal to reduce the TV, graph learning inversely learns the graph structure to minimise the TV [76]. In other words, the TV is used as an optimisable objective function with the graph structure as its argument. Other terms can be added to this objective function to add further constraints on the graph. For example, one can further require the graph to have positive weights and to be sparse, i.d., to have few non-zero weights. While this graph-learning method has connections to learning Markov random fields, GSP adds a new perspective in terms of signals by interpreting the graph learning as minimising the TV [77]. Kalofolias has introduced an algorithm in 2016 to solve this graph learning problem [78], which has since been picked up to retrieve graphs from neurophysiological signals [27, 79, 80]. Section 4.1 explores the link between this TV-based graph retrieval and functional connectivity-based graph retrieval.

4 Links in Graph Signal Processing (GSP)

The plethora of graph retrieval methods and graph representations results in a deep ambiguity in GSP. To begin with, there are several ways to retrieve the graph structure for the data (see Sect. 3.1). For some of the retrieved raw graphs, further preprocessing steps may be possible or required depending on the application, such as setting negative weights to zero, taking the absolute value of the weights, or normalising the graph (see also Sect. 3.2). Lastly, several graph representations can be computed from the preprocessed graphs, such as the adjacency matrix, the Laplacian matrix, or the normalised Laplacian matrix (Sect. 2.4). This multitude of graphs and representations thereof poses the question of whether there are similarities between them that could minimise the ambiguity in GSP. To this end, this subsection investigates the conditions for which some of the graph retrieval methods are equivalent and explores the similarities between different graph representations.

4.1 Links Between Graph Retrieval Methods

This section specifically looks at the links between functional connectivity-based graphs and total variation-optimised graphs.

Kalofolias already showed such a link for a commonly used graph $\mathbf{A}_{\text{exp}} \in \mathbb{R}^{N \times N}$, which is constructed from the rows $\mathbf{x}_{i*}$ of the data matrix $\mathbf{X}$ as follows [78]:

$$a_{ij}^{(\text{exp})} = \exp\left(-\frac{\|\mathbf{x}_{i*} - \mathbf{x}_{j*}\|_2^2}{\sigma}\right).$$

Given the following regularisation term,

$$f^{(\log)}(\mathbf{A}) = \sigma^2 \sum_i \sum_j a_{ij} \left(\log\left(a_{ij}\right) - 1\right),$$

this functional connectivity-based connectivity matrix minimises the edge-based total variation:

$$\underset{\mathbf{A} \in \mathcal{A}}{\operatorname{argmin}}\ \mathrm{TV}_{\mathbf{A}}^{(e)}(\mathbf{X}) + f^{(\log)}(\mathbf{A}) = \mathbf{A}_{\text{exp}},$$

where $\mathcal{A}$ denotes the set of real-valued $N \times N$ adjacency matrices.

In the following, suppose we are given a normalised multivariate signal $\mathbf{X}_{\text{norm}}$, where each row $\mathbf{x}_{i*}$, corresponding to the time signal at sensor i, is standardised to mean 0 and standard deviation 1. We show that for such a signal, a similar link between a total variation-optimised graph and a functional connectivity-based graph, whose entries in the adjacency matrix are the pairwise Pearson correlations, can be established. Note that due to the normalisation of the time signals, this Pearson correlation matrix is equivalent to the covariance matrix.

We firstly introduce the regularisation term

$$f^{(\text{c})}(\mathbf{A}) = \frac{1}{2} \|\mathbf{J}_N - \mathbf{A}\|_F^2 = \frac{1}{2} \sum_i \sum_j (1 - a_{ij})^2, \tag{12}$$

where $\mathbf{J}_N$ is an all-ones matrix of size $N \times N$, and $\| \cdot \|_F$ denotes the Frobenius norm. Specifically, this regularisation term aims to shift the entries a_{ij} of $\mathbf{A}$ closer to 1.

Then, the solution to the optimisation problem

$$\underset{\mathbf{A}\in\mathcal{A}}{\operatorname{argmin}}\ \mathrm{TV}^{(e)}_{\mathbf{A}}(\mathbf{X}_{\text{norm}}) + f^{(\mathrm{c})}(\mathbf{A}) \tag{13}$$

is the adjacency matrix given by the correlation matrix $\mathbf{A}_{\text{corr}}$, where the entries $a^{(c)}_{ij}$ are calculated as the Pearson correlation between the normalised time signals $\mathbf{x}_{i*}$ and $\mathbf{x}_{j*}$ of the sensors i and j, respectively. This can be shown by taking the derivative with respect to a_{ij}:

$$\begin{aligned}
&\frac{\partial}{\partial a_{ij}} \mathrm{TV}^{(e)}_{\mathbf{A}}(\mathbf{X}_{\text{norm}}) + f^{(\mathrm{c})}(\mathbf{A})\Big|_{\mathbf{A}=\mathbf{A}^{\max}} \\
&\qquad = \frac{\partial}{\partial a_{ij}} \sum_{k=1}^{N_t} \sum_{i,j} \left(a_{ij} x_{ik}^2 - a_{ij} x_{ik} x_{jk} + \frac{1}{2}(1 - a_{ij})^2 \right) \Big|_{a_{ij}=a_{ij}^{\max}} \\
&\qquad = \sum_{k=1}^{N_t} \left(x_{ik}^2 - x_{ik} x_{jk} + (1 - a_{ij}^{\max})(-1) \right) \\
&\qquad = \sum_{k=1}^{N_t} \left(1 - x_{ik} x_{jk} - 1 + a_{ij}^{\max} \right) = 0 \\
&\quad \Rightarrow a_{ij}^{\max} = \frac{1}{N_t} \sum_{k=1}^{N_t} x_{ik} x_{jk} = a_{ij}^{(c)}.
\end{aligned}$$

In other words, the correlation matrix $\mathbf{A}_{\text{corr}}$ minimises the total edge-based variation of a normalised data matrix $\mathbf{X}$ given the regularisation term defined in (12). Therefore, it can be equally retrieved by learning the solution to the optimisation problem in (13). Note that the requirement that the signals have to be normalised is naturally given for many neurophysiological signals, such as EEG signals. In our experiment, beginning in Sect. 5, we normalise the data and use the Pearson correlation matrix as the graph for the data.

4.2 *Links Between Graph Representations*

There are arguments for both using the adjacency or the Laplacian matrix as the GSO in GSP, and both are frequently utilised in the literature. In a direct comparison, Huang et al. have observed no noticeable difference between using the adjacency or the Laplacian matrix as the GSO [28]. This observation poses the question of the relation between the use of the adjacency and the Laplacian matrix in GSP. Here, we demonstrate that for large graphs with normally distributed adjacency matrix weights, the two GSOs have similar eigenvectors. This in turn would mean that their respective GFTs are similar.

We here assume that the weights of the adjacency matrix follow a normal distribution with mean μ and standard deviation σ. Accordingly, the entries of the diagonal matrix are given by:

$$d_i = \sum_j a_{ij} = \mu(N-1) + \epsilon_i,$$

where $\epsilon_i \sim \mathcal{N}(0, \sigma^2_{\text{diag}})$, and $\sigma_{\text{diag}} = \sigma\sqrt{N-1}$. Note that the standard deviation σ_{diag} of elements d_i on the diagonal is by a factor of $\sqrt{N-1}$ larger than that of the non-diagonal elements a_{ij}. However, for sufficiently large N, the deviation is significantly smaller than the sum of the expected magnitudes of the non-diagonal elements in that row:

$$\begin{aligned}
\sigma_{\text{diag}} = \sigma\sqrt{N-1} \ll & \sigma\sqrt{\frac{2}{\pi}}(N-1) \\
\leq & \sigma\sqrt{\frac{2}{\pi}}(N-1)e^{-\mu^2/2\sigma^2} \\
& \quad + \mu(N-1)\left(1 - 2\Phi\left(-\frac{\mu}{\sigma}\right)\right) \\
= & \sum_{j \neq i} \mathrm{E}\left[|a_{ij}|\right].
\end{aligned}$$

Here, Φ denotes the normal cumulative distribution function. Consequently, the Laplacian $\mathbf{L}$ can be approximated by a matrix $\mathbf{L}'$, which is defined as follows:

$$\begin{aligned}
\mathbf{L} &= \mathbf{D} - \mathbf{A} \\
&= \mu(N-1)\mathbb{1} + \text{diag}(\epsilon_1, \ldots, \epsilon_N) - \mathbf{A} \\
&\approx \mu(N-1)\mathbb{1} - \mathbf{A} =: \mathbf{L}'.
\end{aligned}$$

Crucially, as eigenvectors $\mathbf{u_k}$ of $\mathbf{A}$ are also eigenvectors of $\mathbf{L}'$,

$$\mathbf{L}'\mathbf{u_k} = \mu(N-1)\mathbf{u_k} - \mathbf{A}\mathbf{u_k} = \lambda'_k\mathbf{u_k},$$

the GFT based on the adjacency matrix and the one based on the Laplacian matrix will be similar.

The normality assumption can be a good approximation for highly interconnected networks with negative weights. The Pearson correlation graphs retrieved from the simulated data in the experimental section of this book chapter fulfil both conditions. For disconnected networks, the assumption is not met as weights between disconnected parts of the graph are zero and thereby not randomly sampled. Considering brain networks, certain pathological conditions, such as split brain, can cause the brain graph to be disconnected. On the other hand, graphs with only positive graph weights can equally violate the normality assumption if the mean is not much larger than the standard deviation, as this gives rise to a skewed distribution. Many graph retrieval methods yield graphs with only positive weights. The topic of negative graph weights was explored in Sect. 3.2.

4.3 Link to Principal Component Analysis (PCA)

PCA projects multivariate samples onto a set of orthogonal components, which are the eigenvectors of the covariance matrix of the data set. Crucially, if the covariance matrix is used for the GFT, then GFT and PCA are mathematically equivalent. Note that for normalised signals with standard deviation 1, which is common for some neurophysiological signals such as EEG signals, the correlation and the covariance matrix are the same. Note also that the diagonal entries of the signal correlation matrix are one and will not affect the eigenvectors. In other words, using the correlation matrix of normalised signals for GFT, either with or without diagonal elements, is equivalent to PCA.

5 Methodology

The aim of the experimental section is to expand on previous work by Rui et al. [62] and Ménoret et al. [27] and investigate dimensionality reduction via GFT projection for classification tasks. The focus will be on understanding the role of graph frequency ordering. To increase the precision needed to compare the graph frequencies, we developed an algorithm to simulate multivariate signals with spatio-temporal features. After transforming the temporal signals into graph frequency signals, our classification model extracts power spectral densities as features from the graph frequency signals and analyses these features using an SVM classifier. We hypothesise that lower graph frequencies amplify spectral features present in the time signals, to the extent that the lower graph frequency signals outperform all other univariate signals tested. On the other hand, higher graph frequency signals, which are based less on the graph structure, should perform equal to or slightly higher than baseline.

Algorithm 1 Neurophysiological signal simulation

Input $N_t, \mathbf{A}_s, h, \alpha, \beta, \gamma$
Output $\mathbf{X}$
1: $\epsilon_{ij}, \tilde{\epsilon}_{ij} \sim \mathcal{N}(0, 1^2), \quad i = 1, \ldots, N_c, j = 1, \ldots, N_t$
2: $\widehat{\boldsymbol{\epsilon}}_{i*} \leftarrow \left(\mathcal{F}^{-1} \circ h \circ \mathcal{F}\right)(\tilde{\boldsymbol{\epsilon}}_{i*}), \quad i = 1, \ldots, N_c$
3: $\mathbf{x}_{*1} \leftarrow \beta \widehat{\boldsymbol{\epsilon}}_{*1}$
4: **for** $t \leftarrow 2, N_t$ **do**
5: $\quad \widehat{\mathbf{x}}_{*t-1} \leftarrow \mathbf{x}_{*t-1} - \bar{\mathbf{x}}_{*t-1} + \gamma \boldsymbol{\epsilon}_{*t}$
6: $\quad \mathbf{x}_{*t} \leftarrow \alpha \mathbf{A}_s \widehat{\mathbf{x}}_{*t-1} + \beta \widehat{\boldsymbol{\epsilon}}_{*t}$
7: **end for**

5.1 Simulated Data Set

We developed Algorithm 1 to generate multivariate signal samples $\mathbf{X}_{sim} \in \mathbb{R}^{N_c \times N_t}$ with N_c channels and N_t time samples. The generated samples simulate neuroimaging measurements, such as EEG, with which they share two crucial characteristics: Firstly, the samples have an underlying graph structure, such that each pair of signals has a specific connectivity associated to it. Secondly, the temporal signals have a specific spectral profile.

The algorithm takes as input the number of generated time samples N_t, the weighted adjacency matrix $\mathbf{A}_s \in \mathbb{R}^{N_c \times N_c}$, the filter function h, as well as parameters α, β and γ. While $\mathbf{A}_s$ controls the connectivity, or spatial structure, the filter function h controls the spectral profile. Specifically, using two filter functions h_1 and h_2 allows to simulate two separate conditions, whereby the similarity between the two filter functions controls the similarity between the conditions. This may simulate conditions such as Alzheimer's disease or healthy control in EEG data, where each condition has a specific spectral profile [81]. Generally, the more similar the conditions are, the harder it becomes to classify the dataset. Parameter α controls the strength of the correlation structure, while parameter β controls the strength of the spectral structure. Parameter γ controls the self-amplification of the signals during the simulation.

In each time step of Algorithm 1, firstly the graph signal of the previous time step $\mathbf{x}_{*t-1}$ is centred around zero and Gaussian noise $\gamma \boldsymbol{\epsilon}_{*t}$ is added to this signal (line 5). Secondly, the adjacency matrix $\mathbf{A}_s$ is used as a GSO to translate the graph signal in time, which enforces the structural connectivity $\mathbf{A}_s$ in the data (line 6). Thirdly, the translated signal is scaled by α and normalised coloured noise $\widehat{\boldsymbol{\epsilon}}_{*t}$ is added scaled with β, whose spectral density profile is controlled by the filter h (line 6). Finally, the simulated multivariate signal is labelled with its condition.

To simulate our data, we used three different filter functions h_2 for condition 2 (orange lines in Fig. 4c), resulting in data sets which are either easy, medium, or difficult to classify. The connectivity structure matrices $\mathbf{A}_s \in \mathbb{R}^{N_c \times N_c}$ with $N_c = 25$ were generated by drawing weights $a^s_{ij} = a^s_{ji} \sim \mathcal{U}_{[-0.1, 0.4]}$ from a uniform distribution for $i > j$, and setting $a^s_{ii} = 0.4$. We set the simulation parameters to

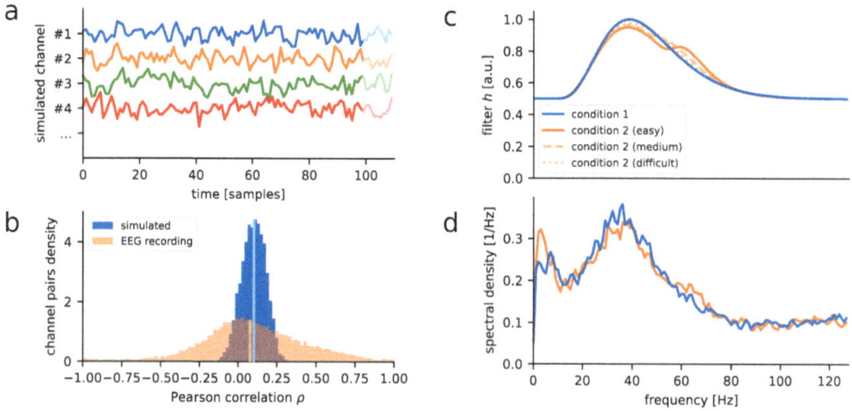

Fig. 4 Simulated neurophysiological signals. (**a**) Time signals for four selected channels across the first 100 samples. Channels #2, #3 and #4 are positively, weakly, and negatively correlated to channel #1, respectively. (**b**) Density of correlations of channel pairs in blue, exhibiting the spatial connectivity structure. In the simulation, this structure is controlled by the matrix $\mathbf{A_s}$, but it is also affected by the simulation parameters. Realistic EEG recordings (light orange) exhibit a wider spread of correlations, which is in part due to volume conduction effects not included in the simulation. The median of the simulation (vertical line) was matched to that of the EEG recording. (**c**, **d**) Demonstration of power spectral density control and adjustment of classification difficulty. (**c**) The Fourier filter function h used to colour the noise in Algorithm 1, assuming a sampling frequency of 256 Hz. For each data set, two conditions are simulated. Condition 2 can be varied to make it easy (solid), medium (dashed), or difficult (dotted) to distinguish from condition 1. The difficulty depends on the similarity between the two conditions. (**d**) Welch power spectral density of simulated signals averaged across all channels for two easily distinguishable conditions. Figures (**c**) and (**d**) clearly show that the shape of the power spectral density profile of the simulated signals can be controlled. Parameter α in Algorithm 1 can be used to reduce the power density at lower frequencies

$\alpha = 0.5$, $\beta = 1$, and $\gamma = 1$. For each simulated participant, we generated $N_c = 25$ time-series signals with $N_t = 2048$ time samples each.

Figure 4 shows the simulated signals (a) along with their spatial (b) and spectral (d) structure. The spatial structure in (b) is shown as the density of channel pairs at a specific correlation value, together with the spatial structure of real EEG recordings (orange, for a description of the dataset see Sect. 3.3). The median of the simulated data was matched to that of the EEG recording using the parameters in the algorithm. The spread of the correlation values in EEG recordings is higher, which may be partly due to volume conduction effects not considered in this simulation [82]. However, these artefacts can be principally reduced by transforming the EEG signals from the sensor space to the source space using sophisticated source reconstruction methods [83], which also allows to recover the true underlying connectivity structure more accurately [84, 85]. The spectral structure (D) of the simulated signals is enforced by the spectral density profile of the coloured noise (C), demonstrating that the spectral structure can be controlled.

5.2 Classification Model

The goal of our classification model is to GFT-transform the multivariate signal into graph frequency signals and subsequently classify the samples into the two conditions, using only one graph frequency signal at a time. These graph frequency signals may capture spectral features present in multiple time signals, or even spatio-temporal features not present in isolated one-dimensional time signals. The structure of our graph frequency analysis is illustrated in Fig. 5. The approach allows us to determine the classification performance for each graph frequency, which we used to assess the expressiveness of the spectral structures in this graph frequency signal.

The first step of the analysis consists of retrieving the graph structure from the training samples, for which we used the functional connectivity. Note that we need to use the same graph for all samples to keep the graph Fourier modes constant. Specifically, we computed the correlation matrix for each sample in the training set and subsequently averaged all matrices, yielding a common weighted adjacency matrix. Secondly, we computed the GFT matrix and carried out the GFT, yielding $N_c = 25$ graph frequency signals. We used the weighted adjacency and the Laplacian matrix for the GFT in two separate experiments. Thirdly, we extracted spectral features from those signals for each sample. To this end, we computed the Welch power spectral density with a window of 32 time samples for each transformed signal and selected the first $N_f = 20$ power spectral densities as our features. Assuming a sampling frequency of 256 Hz, the extracted spectral

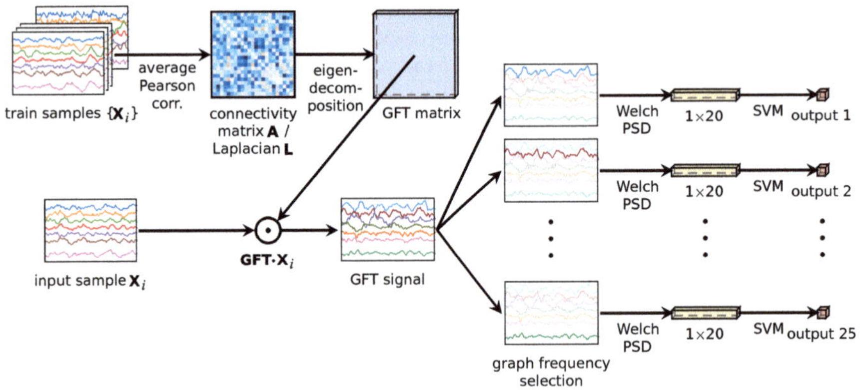

Fig. 5 Illustration of the graph frequency analysis. For each cross-validation iteration, all simulated samples from the training set are used to construct either the connectivity matrix **A** or the Laplacian matrix **L**, from which the GFT matrix is computed as the eigendecomposition (see Sect. 2.4). Using the GFT matrix, the input sample $\mathbf{X}_i$ is transformed to its GFT signal $\tilde{\mathbf{X}}_i$. The signal is further split into the 25 graph frequency signals. Lastly, a support vector machine classifier is trained on the 20 time spectral features extracted from each graph frequency signal using Welch's power spectral density method. The performance of each graph frequency can then be used to assess the quality of the spectral features in the graph frequency signal

densities cover frequencies in the range of 0–76 Hz at a resolution of 4 Hz. Lastly, we trained an SVM classifier separately for each graph frequency to classify the labelled samples, using only the 20 features calculated from each graph frequency signal. Specifically, we used a common SVM configuration consisting of a non-linear radial basis function kernel with kernel coefficient $\gamma = 1/(N_f \operatorname{Var}(X))$, where X is the feature vector, and a regularisation strength of $C = 1$. Note that we did not optimise these hyperparameters for the following two reasons: Firstly, our study was focused on model comparison rather than performance optimisation. Secondly, the feature extraction in our model is fixed, meaning that hyperparameter configuration biases are unlikely. The baseline classification models, to be introduced in Sect. 5.4, follow the same analysis steps and differ from this main model only in that their graph structure is modified.

5.3 Testing

Although our method allows to arbitrarily generate data, this process is still time-consuming: On the one hand, each time sample in a simulated participant is generated from the previous sample, restricting the use of vectorised computations. On the other hand, large numbers of samples on the order of tens of thousands are required to achieve the precision needed to compare individual graph frequencies. We therefore introduce a modified, perpetual version of cross-validation, as illustrated in Fig. 6, which to the best of our knowledge has not been used before. The main goal is to successively generate unlimited pairs of training and test sets, with the constraint that each generated sample is only used once for testing. Initially, $N_s = 100$ samples are generated, divided evenly in condition 1 and condition 2. This data set is split into $k = 10$ folds. Nine of these are used for training, resulting in 90 training samples, while the remaining 10 samples are used for testing. For the subsequent iteration, data are generated to form a new fold with $N_s/k = 10$ samples, which is added to the training set, whereas the testing fold is shifted by one

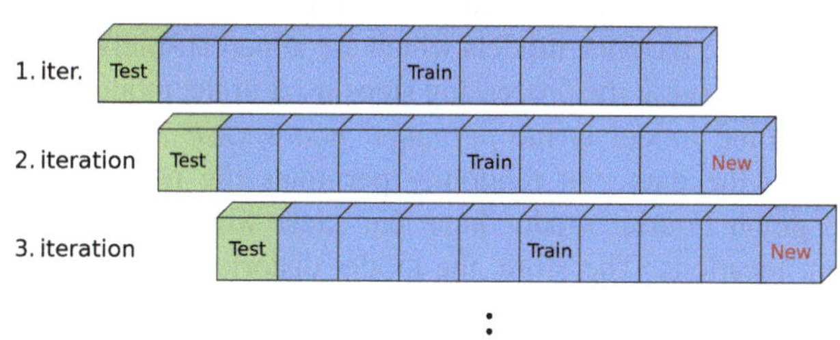

Fig. 6 Illustration of the perpetual cross-validation used during the analysis. The whole data set is split into $k = 10$ folds. One fold is assigned to be the testing set, while the remaining $k - 1$ folds comprise the training set. For the subsequent iteration, a fold with newly generated data is added to the training set, while the testing fold is shifted by one fold

fold as illustrated in Fig. 6. Overwriting the discarded fold with the new fold ensures that only $N_s = 100$ samples have to be stored in memory, or $N_s + N_s(1-1/k) = 190$ samples if the initially generated samples are reused during the last iterations. The final accuracy scores are averaged across all samples in the testing set and across all iterations.

The sample size of the training set is chosen to mimic the number of recorded samples in realistic neuroimaging data sets, such as the sparse data set mentioned in Sect. 3.3. For each simulated graph structure, we ran 90 iterations, excluding the initial configuration, yielding 1000 generated samples, which is equal to the number of testing samples. We simulated 40 graph structures for each difficulty level, resulting in overall 120000 generated samples, each of which is a data matrix of size 25×2048. While matching the training size to real-life experiments ensures that the performance values of the simulated experiment are similar to those of the experiment, repeating these simulations arbitrarily often allows to increase the precision of the retrieved performance to the required level.

5.4 Baseline Models

The principal goal of the baseline framework is to assess the gain of using GSP in our data classification procedure. Crucially, such a framework must be able to pinpoint the performance gain to the graph structure in the data and not incidental gains due to other aspects of the methodology. To illustrate this point, we firstly discuss the baseline model in Huang et al. [28]: Here, surrogate signals are constructed for the baseline model by GFT-transforming the signals, then randomly flipping half of the signs of the Fourier components, before finally performing the inverse GFT. Their model is based on graph band-pass filtering the signals; however, the same operation was not performed on the surrogate signals. In other words, not only the graph but also operations carried out on the signals differ between model and baseline. As a second illustration, Menorét et al. used graph-free models such as PCA as their baseline model [27]; however, such baselines principally differ from the tested model in more ways than just the graph structure. In contrast, here we compose a set of three baseline models which are aimed at solely altering the graph structure, while preserving the processing steps used in the model.

The first baseline model is the *permuted nodes GSP model*, which uses the graph underlying the data, but randomly permutes the nodes of this graph. In this way, the graph Fourier modes have the same weights as the actual model, but at different locations. Therefore, the model distorts the graph structure, while retaining important characteristics of the modes such as the sparsity or diversity. Figure 7 illustrates the effect of this permutation on the modes in relation to the graph structure. While the top row (a–e) shows the lowest and highest modes on the graph, the bottom row (f–j) shows the modes for the permuted graph for the same graph frequencies. The graph Fourier modes clearly characterise the graph structure at lower graph frequencies for the unaltered graph. However, any such

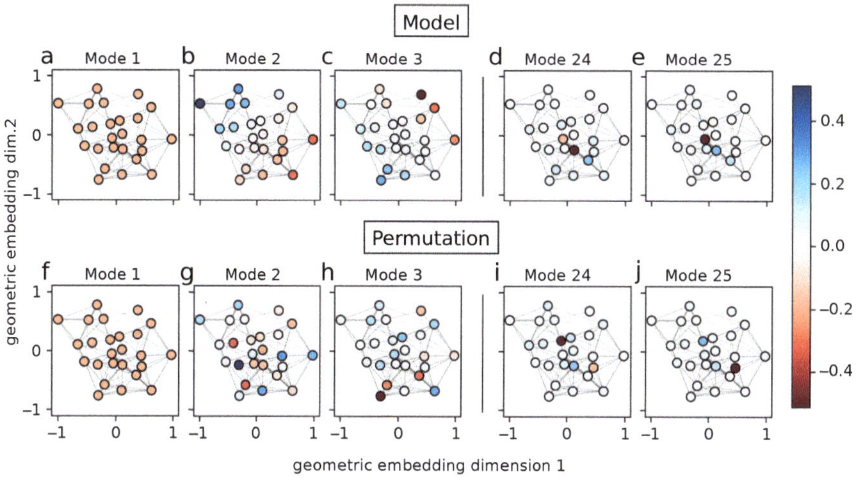

Fig. 7 Retrieved functional connectivity graph of a typical simulated training data set, visualised as a geometric graph in two dimensions. The geometric positions were determined with the Fruchterman–Reingold force-directed algorithm using the NetworkX package in Python [86]. Node colourings represent the graph Fourier modes computed from the graph Laplacian. The top row (**a**–**e**) shows the three lowest (modes 1–3) and highest (modes 24 and 25) graph Fourier modes for this graph. While the low-frequency modes clearly exhibit a wave-like pattern covering the whole graph, a similar pattern is not visible for the higher modes. The bottom row (**f**–**j**) shows the same modes for a permuted graph, which is used in the permuted nodes baseline model. The permutation distorts any low-frequency ordering on the graph, while not changing the sparsity of the modes. Mode 1 (**a**,**f**) corresponds to the DC mode and is not affected by the permutation

relation between the graph and the modes is distorted for the permuted graph. We expect this baseline model to perform equal to or worse than our classification model across all graph frequencies, given that the signal transformation in this model is not based on the spatial structure. Out of the three baseline models, this model is closest to the original classification model.

The second, related baseline model is the *random graph GSP model*, which uses GSP with a randomly generated graph. Specifically, we generated an undirected graph without self-loops and normally distributed weights $w_{ij} = w_{ji} \sim \mathcal{N}(1, 1^2)$, such that roughly 84 % of the weights are positive. While this model transforms the data using GFT, these transformations are not based on the actual graph structure. Similarly to the first baseline model, we do not expect this model to outperform the original model for any of the graph frequencies due to the random nature of the transformation.

The third baseline model is the *single channel model*, which does not transform the data and is equivalent to conventional signal processing. It can also be viewed as GSP with the identity transform, given algebraically by the identity matrix $\mathbb{1}$. Following this notion, the "graph frequency signals" are given by the single channels and have no natural ordering. This trivially means that the mean performance is the same across all channels. Note also that the "graph Fourier modes" are vectors which are 1 at the channel index, and 0 elsewhere, which means that these modes are maximally sparse. This model is the only model that excludes the graph structure, making it an indispensable baseline model.

6 Results

Figure 8 illustrates the results of our analysis of the artificially generated data. Specifically, it shows the classification accuracy in dependence on the graph frequency of the transformed signal for the three simulated difficulty levels. The

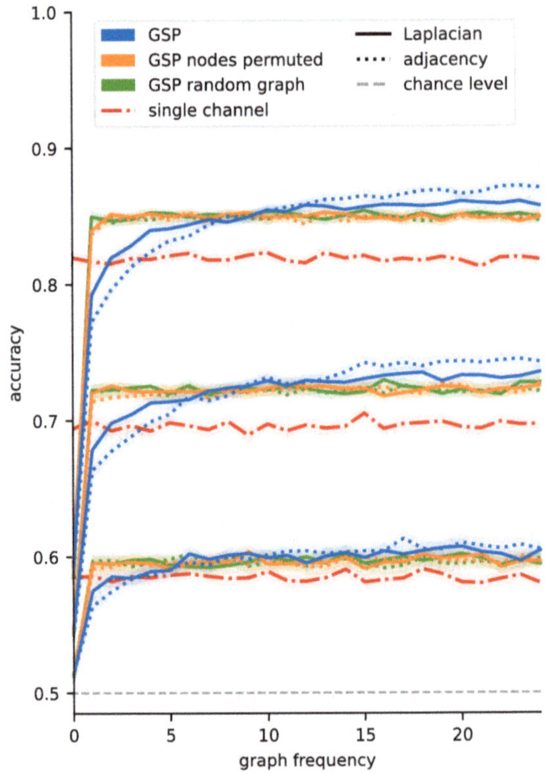

Fig. 8 Classification accuracy as a function of the graph frequency of the transformed signals. Three data sets with varying classification difficulty were simulated, whereby the difficulty was controlled by modifying the filter h as shown in Fig. 4c. Easy, medium, and difficult classification difficulties result in the high, medium, and low accuracies overlaid in the figure, respectively. The model is shown in blue, while the baseline models are shown in orange, green, and red. Models using the graph Fourier transform with the Laplacian (adjacency) matrix are shown in solid (dotted) lines. Note that the single channel baseline model is not based on Graph Signal Processing (GSP) and thereby does not make use of the adjacency or the Laplacian matrix. This also means that its univariate signals are not ordered by graph frequency, but by graph node

main model, shown in blue, is compared against the three baseline models. The accuracy scores in the figures are averaged across all testing samples, as described in Sect. 5.3.

To begin with, the model performance increases steadily with increasing graph frequency, contrary to our expectations. This effect is even more pronounced in the adjacency-based model, which increases at a higher rate. The effect does not seem to depend on the difficulty level, albeit being more pronounced at the easier difficulty level. Crucially, the same effect is not present in the graph-based baseline models, whose performance is constant across all graph frequencies, apart from the graph DC frequency. While the two baselines outperform the model at lower frequencies, their accuracy is slightly lower than the model at higher frequencies. Differences between the two baselines are not significant and seem to be negligible. The accuracy of the single channel model, on the other hand, is trivially constant across all channels. As expected, the accuracy is mostly lower than that of all other models. The accuracy of the lowest graph frequency signal, namely the graph DC component, is by far the lowest for the three graph-based models. Note that the graph Fourier mode corresponding to the graph DC component is constant (Laplacian GFT) or near constant (adjacency GFT) across all entries and thereby highly dense.

Table 1 shows the mean accuracy at high graph frequencies of each model. Only the easiest simulated difficulty level was considered, as the effects are most pronounced at this level. To compute the mean, the five highest graph frequencies were averaged, given that all models have plateaued in this range. The adjacency-based main model leads the table significantly with 87.0% accuracy, outperforming the Laplacian-based counterpart by 1% point. The two graph-based baseline models have an accuracy of roughly 85%, irrespective of the graph representation used. The single channel baseline model trails all other models, performing more than 5% points lower than the best-performing model. These quantitative results indicate that our model significantly outperforms the baseline models at high graph frequencies.

Table 1 Mean accuracy of all tested models for the easiest simulated difficulty level, averaged across the highest five graph frequencies. The best-performing model (marked in bold) is the main model based on Graph Signal Processing (GSP) using the adjacency matrix, followed by the main model using the Laplacian matrix. The model with the lowest accuracy is the single channel baseline model, which is more than 5% points lower than the best-performing model

Model	Graph representation	Mean accuracy [%] (5 highest frequ.)	CI [%]
GSP	Laplacian	86.0	[85.8,86.1]
	Adjacency	**87.0**	[86.8,87.1]
GSP (nodes permuted)	Laplacian	84.9	[84.7,85.0]
	Adjacency	84.9	[84.8,85.1]
GSP (random graph)	Laplacian	85.1	[84.9,85.2]
	Adjacency	84.9	[84.7,85.0]
Single channel	–	81.7	[81.6,81.9]

7 Discussion

In the experiment, we observed the effect that the performance of the classification model increases with increasing frequency of the graph frequency signal. This suggests that projections on low-frequency graph Fourier modes, which mix nearby channels, do not amplify spectral features. Conversely, projections on high-frequency modes, which mix channels from distant parts of the graph, seem to accumulate spectral features more effectively. We hypothesise that the spectral features in closely connected channels are similar, because spectral features may spread to nearby locations during the data generation. Features from distant channels, on the other hand, are better isolated from one another, allowing them to contribute more diverse spectral features to the graph frequency signals. The graph-based baseline models are in line with our hypothesis. Both models solely differ from the main model in that they use a graph unrelated to the data. This means that projections on all non-DC modes mix channels from random location, as shown in Fig. 7. However, these patterns effectively correspond to mid-frequency modes on the actual graph across all non-DC graph frequencies, which means that their performance matches those of mid-frequency modes and does not change with graph frequency.

We further analyse effects due to the sparsity of the graph Fourier mode. To begin with, the eigenmodes in the single channel baseline model are simply the rows of the identity matrix and thereby maximally sparse. As expected, the performance of this model is lower than that of all graph-based models at higher graph frequencies, possibly because these modes do not source information from more than one channel, that is, their receptive field is minimal. The graph DC Fourier mode, on the other hand, is the mode which is maximally dense. Even though its receptive field covers all channels, its performance is nevertheless the lowest by far. We attribute the poor performance of highly dense modes to interference effects: If too many channels are mixed together, spectral features interfere with each other due to phase differences between the channels [87], thereby attenuating these spectral features. Therefore, the graph Fourier modes with the highest performance are those which are neither too sparse nor too dense. Importantly, the permutation model baseline allows us to exclude that the main effect observed in the model, namely the increase in performance with increasing graph frequency, is due to the sparsity of the modes: While the sparsity of the permuted eigenmodes is the same as those of the model, the main effect is not present in the permutation model.

Taken together, there seem to be two mechanisms at play that explain the performance of the graph frequency signals. Firstly, the performance depends on how many spectral features are accumulated during the transformation. While nearby channels share spectral features, distant channels have different spectral features. Generally, only transformations that are based on the actual graph can selectively mix nearby or distant channels. Secondly, the more channels are mixed together, that is, the denser the graph Fourier mode is, the more the channels interfere with each other. These two mechanisms explain our main result, namely the superior performance of high-frequency graph frequency signals, as follows: Projections on high-frequency graph Fourier modes mix only few, but distant channels together; hence, they optimally accumulate diverse spectral features, while keeping interference effects in check.

Our observation that high-frequency graph frequency signals outperform both low-frequency graph frequency signals as well as unaltered single channels may prove useful for GSP-based applications such as dimensionality reduction or pruning of neural networks. However, note that the focus of our study was on understanding the specific role of individual graph frequencies. Hence, our study did not explicitly test and compare using a range of graph frequency signals for classification, which would almost certainly increase accuracy and thereby prove more useful for applications. An alternative approach to dimensionality reduction is to reduce the dimensionality of the data in the feature domain, instead of the signal domain, which has the advantage that interference effects are avoided. Our method may be repurposed for feature selection by limiting the spectral features to those of the most relevant channels, such as the channels that contribute most to the high-frequency graph frequency signals.

Interestingly, the adjacency-based model outperforms the Laplacian-based model slightly at higher graph frequencies. Following our interpretation of the results, this suggests that projections on high-frequency adjacency-based graph Fourier modes may sample more distant parts of the graph than their Laplacian-based counterparts. Generally, however, the choice of the graph representation for the GFT does not have a strong impact on the results, which is in line with previous observations by Huang et al. [28].

The superior performance of high-frequency over low-frequency graph frequency signals aligns with results obtained by Ménoret et al. on fMRI data, who observed the same effect for five out of the seven graphs they used, including a correlation-based graph [27]. However, their results are contradictory: They also found that this effect was reversed for a smooth graph optimised by minimising the TV with the algorithm by Kalofolias [78]; however, we showed in Sect. 4.1 that TV-optimised graphs share crucial characteristics with correlation-based graphs. On the other hand, our main result contradicts results obtained by Ouahidi et al., who found that keeping low-frequency graph frequency signals leads to the highest classification accuracy on fMRI data [36]. We attribute this discrepancy to differences in the spectral and spatial structure between neurophysiological signals such as EEG or MEG on the one hand, which we simulated in our experiment, and fMRI signals on the other. Firstly, EEG and MEG signals have arguably a richer spectral structure compared to fMRI signals, which means that interference effects play a much greater role in those signals. Secondly, fMRI data have a richer spatial structure: To illustrate this, Ouahidi et al. extracted a graph with 360 nodes from their data set, while our simulation was limited to a multivariate signal with 25 nodes. Both differences are likely to have an effect on the two proposed mechanisms explaining our results.

Our simulated data may differ from real EEG or MEG data in crucial ways. To begin with, the spatial structure of our simulated data is not very pronounced, i.d., the pairwise correlation-based functional connectivities does not vary much. In contrast, the spatial structure in real-life EEG data is much more pronounced, which may partly be the result of volume conduction effects. Additionally, the spectral structure may differ from real data as well. While the spectral structure in our simulation is enforced by adding coloured noise, the spectral structure in real-life neuroimaging data may rely more strongly on the network, forming spatio-temporal structures not present in our simulation. Nevertheless, we argue that the generated data capture the essential characterisics of neurophysiological signals relevant for GSP, as shown in Fig. 4. Hence, their analysis advocates the use of GSP for real-life neurophysiological EEG signals and gives valuable insight into possible GSP mechanisms.

8 Conclusion

In this book chapter, we have derived central GSP concepts theoretically, reviewed selected applications of GSP, and investigated the meaning of graph frequency in GSP experimentally, with applications to neuroimaging in mind. In our view, one of the most significant challenges that GSP is currently facing is the ambiguity of the GFT, which is due to the number of choices for how to retrieve the graph (see Sect. 3.1) and for which graph representation to use. To this end, we put the focus of this book chapter on understanding, validating, and interpreting GSP concepts, rather than performance optimisation.

The theoretical section of this book chapter aimed to give the reader a comprehensive introduction into GSP concepts and how they are derived. Importantly, we juxtaposed the alternative formulations of the GFT and investigated similarities between these formulations, which may guide researchers in selecting and interpreting a specific formulation. Lastly, we explored links to established methods such as PCA, allowing the reader to place GSP in a wider context.

From an empirical perspective, the ambiguity of the GFT highlights the need for a thorough validation procedure. To this end, we have introduced a baseline testing framework for GSP in Sect. 5.4, consisting of a set of overall three baseline models. For our experiment, we have developed a simulation algorithm to generate arbitrary amounts of labelled multivariate signals. Using the simulated data, we have classified the data sets by extracting spectral features from graph frequency signals and subsequently feeding the features from each graph frequency signal into an SVM classifier. Our results link the classification accuracy to the graph frequency, thereby enabling us to make inferences about the expressiveness of the spectral features at each graph frequency. Specifically, we obtained the result that high-frequency graph frequency signals outperform both low-frequency signals and all baseline models. The accuracy gradient between low-frequency and high-frequency signals may be leveraged for dimensionality reduction or for pruning graph neural networks. Future work may be directed at testing dimensionality reduction for classification on real-life neurophysiological data sets. To use this method on EEG data, more sophisticated functional connectivity retrieval methods may be required, as well as complex source-space reconstruction methods.

Generating the data proved useful for two reasons: On the one hand, it allowed us to increase the precision of our results arbitrarily, laying bare the underlying GSP mechanisms. On the other hand, knowing how spectral features spread through the network narrowed down possible interpretations of the results. Likewise, the baseline framework proved valuable, as it enabled us to pinpoint the measured effect to the graph structure in the data. Together, our experiment showcased how a combination of simulating the data and a strong validation procedure can propel our understanding of the mechanisms behind GSP. We believe that the mechanisms uncovered here may ultimately inform the design of GSP-based applications for neurophysiological signal processing.

Our study only tested simulated EEG-like multivariate signals. Further studies may focus on simulating more realistic EEG data sets, as wells as simulating other imaging modalitites such as fMRI, in order to find precise and modality-specific mechanisms in GSP. Lastly, future work may also focus on gaining an understanding of mechanisms underlying other GSP applications, such as graph filtering.

Acknowledgments The authors would like to thank Dominik Klepl for important discussions.

References

1. W. Bourgeois, A.-C. Romain, J. Nicolas, and R. M. Stuetz, "The use of sensor arrays for environmental monitoring: interests and limitations," *Journal of Environmental Monitoring*, vol. 5, no. 6, pp. 852–860, 2003.
2. M. T. Vlaardingerbroek and J. A. Boer, *Magnetic resonance imaging: theory and practice*. Springer Science & Business Media, 2013.
3. M. Tubaishat and S. Madria, "Sensor networks: an overview," *IEEE potentials*, vol. 22, no. 2, pp. 20–23, 2003.
4. P. Sawangjai, S. Hompoonsup, P. Leelaarporn, S. Kongwudhikunakorn, and T. Wilaiprasitporn, "Consumer grade EEG measuring sensors as research tools: A review," *IEEE Sensors Journal*, vol. 20, no. 8, pp. 3996–4024, 2019.
5. L. Stankovic, D. P. Mandic, M. Dakovic, I. Kisil, E. Sejdic, and A. G. Constantinides, "Understanding the basis of graph signal processing via an intuitive example-driven approach [lecture notes]," *IEEE Signal Processing Magazine*, vol. 36, no. 6, pp. 133–145, 2019.
6. R. Lindsey and L. Dahlman, "Climate change: Global temperature," *Climate. gov*, vol. 16, 2020.
7. T. Kurokawa, T. Oki, and H. Nagao, "Multi-dimensional graph fourier transform," *arXiv preprint arXiv:1712.07811*, 2017.
8. B. Upadhyaya, M. Kitamura, and T. Kerlin, "Multivariate signal analysis algorithms for process monitoring and parameter estimation in nuclear reactors," *Annals of Nuclear Energy*, vol. 7, no. 1, pp. 1–11, 1980.
9. A. Biasiucci, B. Franceschiello, and M. M. Murray, "Electroencephalography," *Current Biology*, vol. 29, no. 3, pp. R80–R85, 2019.
10. K. J. Miller, P. Shenoy, J. W. Miller, R. P. Rao, J. G. Ojemann *et al.*, "Real-time functional brain mapping using electrocorticography," *Neuroimage*, vol. 37, no. 2, pp. 504–507, 2007.

11. J. Vrba and S. E. Robinson, "Signal processing in magnetoencephalography," *Methods*, vol. 25, no. 2, pp. 249–271, 2001.
12. G. H. Glover, "Overview of functional magnetic resonance imaging," *Neurosurgery Clinics*, vol. 22, no. 2, pp. 133–139, 2011.
13. D. P. Subha, P. K. Joseph, R. Acharya U, and C. M. Lim, "EEG signal analysis: a survey," *Journal of medical systems*, vol. 34, pp. 195–212, 2010.
14. P. A. Abhang, B. W. Gawali, and S. C. Mehrotra, *Introduction to EEG-and speech-based emotion recognition*. Academic Press, 2016.
15. A. Horvath, A. Szucs, G. Csukly, A. Sakovics, G. Stefanics, and A. Kamondi, "Eeg and erp biomarkers of alzheimer's disease: a critical review." *Frontiers in bioscience (Landmark edition)*, vol. 23, pp. 183–220, 2018.
16. R. Lin, R.-G. Lee, C.-L. Tseng, H.-K. Zhou, C.-F. Chao, and J.-A. Jiang, "A new approach for identifying sleep apnea syndrome using wavelet transform and neural networks," *Biomedical Engineering: Applications, Basis and Communications*, vol. 18, no. 03, pp. 138–143, 2006.
17. A. Al-Ezzi, N. Kamel, I. Faye, and E. Gunaseli, "Review of EEG, ERP, and brain connectivity estimators as predictive biomarkers of social anxiety disorder," *Frontiers in psychology*, vol. 11, p. 730, 2020.
18. G. Cheron, G. Petit, J. Cheron, A. Leroy, A. Cebolla, C. Cevallos, M. Petieau, T. Hoellinger, D. Zarka, A.-M. Clarinval *et al.*, "Brain oscillations in sport: toward EEG biomarkers of performance," *Frontiers in psychology*, vol. 7, p. 246, 2016.
19. B. Swiderski, S. Osowski, and A. Rysz, "Lyapunov exponent of EEG signal for epileptic seizure characterization," *Proceedings of the 2005 European Conference on Circuit Theory and Design, 2005.*, vol. 2. IEEE, 2005, pp. II–153.
20. S. J. Luck, *An introduction to the event-related potential technique*. MIT press, 2014.
21. M. R. M. George, G. Potts, D. Kothman, L. Martin, and C. Mukundan, "Frontal deficits in alcoholism: an ERP study," *Brain and cognition*, vol. 54, no. 3, pp. 245–247, 2004.
22. W. S. Pritchard, "Cognitive event-related potential correlates of schizophrenia." *Psychological bulletin*, vol. 100, no. 1, p. 43, 1986.
23. J. M. Peters, M. Taquet, C. Vega, S. S. Jeste, I. S. Fernández, J. Tan, C. A. Nelson, M. Sahin, and S. K. Warfield, "Brain functional networks in syndromic and non-syndromic autism: a graph theoretical study of EEG connectivity," *BMC medicine*, vol. 11, no. 1, pp. 1–16, 2013.
24. F. de Vico Fallani, J. Richiardi, M. Chavez, and S. Achard, "Graph analysis of functional brain networks: practical issues in translational neuroscience," *Philosophical Transactions of the Royal Society B: Biological Sciences*, vol. 369, no. 1653, p. 20130521, 2014.
25. A. Ortega, P. Frossard, J. Kovačević, J. M. Moura, and P. Vandergheynst, "Graph signal processing: Overview, challenges, and applications," *Proceedings of the IEEE*, vol. 106, no. 5, pp. 808–828, 2018.
26. X. Dong, D. Thanou, L. Toni, M. Bronstein, and P. Frossard, "Graph signal processing for machine learning: A review and new perspectives," *IEEE Signal processing magazine*, vol. 37, no. 6, pp. 117–127, 2020.
27. M. Ménoret, N. Farrugia, B. Pasdeloup, and V. Gripon, "Evaluating graph signal processing for neuroimaging through classification and dimensionality reduction," *2017 IEEE Global Conference on Signal and Information Processing (GlobalSIP)*. IEEE, 2017, pp. 618–622.
28. W. Huang, T. A. Bolton, J. D. Medaglia, D. S. Bassett, A. Ribeiro, and D. Van De Ville, "A graph signal processing perspective on functional brain imaging," *Proceedings of the IEEE*, vol. 106, no. 5, pp. 868–885, 2018.
29. S. Itani and D. Thanou, "A graph signal processing framework for the classification of temporal brain data," *2020 28th European Signal Processing Conference (EUSIPCO)*. IEEE, 2021, pp. 1180–1184.
30. H. Behjat and M. Larsson, "Spectral characterization of functional MRI data on voxel-resolution cortical graphs," *2020 IEEE 17th International Symposium on Biomedical Imaging (ISBI)*. IEEE, 2020, pp. 558–562.
31. S. Mortaheb, J. Annen, C. Chatelle, H. Cassol, G. Martens, A. Thibaut, O. Gosseries, and S. Laureys, "A graph signal processing approach to study high density EEG signals in patients

with disorders of consciousness," *2019 41st Annual International Conference of the IEEE Engineering in Medicine and Biology Society (EMBC)*. IEEE, 2019, pp. 4549–4553.
32. S. S. Saboksayr, G. Mateos, and M. Cetin, "EEG-based emotion classification using graph signal processing," *ICASSP 2021–2021 IEEE International Conference on Acoustics, Speech and Signal Processing (ICASSP)*. IEEE, 2021, pp. 1065–1069.
33. K. Glomb, J. R. Queralt, D. Pascucci, M. Defferrard, S. Tourbier, M. Carboni, M. Rubega, S. Vulliemoz, G. Plomp, and P. Hagmann, "Connectome spectral analysis to track EEG task dynamics on a subsecond scale," *NeuroImage*, vol. 221, p. 117137, 2020.
34. D. I. Shuman, S. K. Narang, P. Frossard, A. Ortega, and P. Vandergheynst, "The emerging field of signal processing on graphs: Extending high-dimensional data analysis to networks and other irregular domains," *IEEE signal processing magazine*, vol. 30, no. 3, pp. 83–98, 2013.
35. D. Thanou, X. Dong, D. Kressner, and P. Frossard, "Learning heat diffusion graphs," *IEEE Transactions on Signal and Information Processing over Networks*, vol. 3, no. 3, pp. 484–499, 2017.
36. Y. El Ouahidi, H. Tessier, G. Lioi, N. Farrugia, B. Pasdeloup, and V. Gripon, "Pruning graph convolutional networks to select meaningful graph frequencies for fMRI decoding," *2022 30th European Signal Processing Conference (EUSIPCO)*. IEEE, 2022, pp. 937–941.
37. A. Sandryhaila and J. M. Moura, "Discrete signal processing on graphs," *IEEE transactions on signal processing*, vol. 61, no. 7, pp. 1644–1656, 2013.
38. A. Sandryhaila and J. M. Moura, "Discrete signal processing on graphs: Frequency analysis," *IEEE Transactions on Signal Processing*, vol. 62, no. 12, pp. 3042–3054, 2014.
39. D. K. Hammond, P. Vandergheynst, and R. Gribonval, "Wavelets on graphs via spectral graph theory," *Applied and Computational Harmonic Analysis*, vol. 30, no. 2, pp. 129–150, 2011.
40. A. Sandryhaila and J. M. Moura, "Discrete signal processing on graphs: Graph fourier transform," *2013 IEEE International Conference on Acoustics, Speech and Signal Processing*. IEEE, 2013, pp. 6167–6170.
41. A. Gavili and X.-P. Zhang, "On the shift operator, graph frequency, and optimal filtering in graph signal processing," *IEEE Transactions on Signal Processing*, vol. 65, no. 23, pp. 6303–6318, 2017.
42. M. Defferrard, X. Bresson, and P. Vandergheynst, "Convolutional neural networks on graphs with fast localized spectral filtering," *Advances in neural information processing systems*, vol. 29, 2016.
43. T. N. Kipf and M. Welling, "Semi-supervised classification with graph convolutional networks," *ICLR*, 2017.
44. Q. Li, Z. Han, and X.-M. Wu, "Deeper insights into graph convolutional networks for semi-supervised learning," *Proceedings of the AAAI conference on artificial intelligence*, vol. 32, no. 1, 2018.
45. S. Zhang, H. Tong, J. Xu, and R. Maciejewski, "Graph convolutional networks: a comprehensive review," *Computational Social Networks*, vol. 6, no. 1, pp. 1–23, 2019.
46. J. Shi and J. M. Moura, "Graph signal processing: Dualizing GSP sampling in the vertex and spectral domains," *IEEE Transactions on Signal Processing*, vol. 70, pp. 2883–2898, 2022.
47. S. Chen, A. Sandryhaila, J. M. Moura, and J. Kovacevic, "Signal denoising on graphs via graph filtering," *2014 IEEE Global Conference on Signal and Information Processing (GlobalSIP)*. IEEE, 2014, pp. 872–876.
48. K. J. Friston, C. D. Frith, P. F. Liddle, and R. S. Frackowiak, "Functional connectivity: the principal-component analysis of large (pet) data sets," *Journal of Cerebral Blood Flow & Metabolism*, vol. 13, no. 1, pp. 5–14, 1993.
49. F. He and Y. Yang, "Nonlinear system identification of neural systems from neurophysiological signals," *Neuroscience*, vol. 458, pp. 213–228, 2021.
50. J. S. Damoiseaux and M. D. Greicius, "Greater than the sum of its parts: a review of studies combining structural connectivity and resting-state functional connectivity," *Brain structure and function*, vol. 213, no. 6, pp. 525–533, 2009.

51. Y. Li and G. Mateos, "Learning similarity-preserving representations of brain structure-function coupling," *2022 30th European Signal Processing Conference (EUSIPCO)*. IEEE, 2022, pp. 922–926.
52. X.-J. Wang and H. Kennedy, "Brain structure and dynamics across scales: in search of rules," *Current opinion in neurobiology*, vol. 37, pp. 92–98, 2016.
53. B. Horwitz, "The elusive concept of brain connectivity," *Neuroimage*, vol. 19, no. 2, pp. 466–470, 2003.
54. M. Fraschini, S. M. Pani, L. Didaci, and G. L. Marcialis, "Robustness of functional connectivity metrics for EEG-based personal identification over task-induced intra-class and inter-class variations," *Pattern Recognition Letters*, vol. 125, pp. 49–54, 2019.
55. T. Cattai, G. Scarano, M.-C. Corsi, F. D. V. Fallani, and S. Colonnese, "EEG as signal on graph: a multilayer network model for BCI applications," *2022 30th European Signal Processing Conference (EUSIPCO)*. IEEE, 2022, pp. 932–936.
56. O. Sporns, "Structure and function of complex brain networks," *Dialogues in clinical neuroscience*, 2022.
57. R. Sedgewick, *Algorithms in java, part 5: Graph algorithms*. Addison-Wesley Professional, 2003.
58. Q. Wang, M. A. Powell, A. Geisa, E. Bridgeford, and J. T. Vogelstein, "Why do networks have inhibitory/negative connections?" *arXiv preprint arXiv:2208.03211*, 2022.
59. A. Chambolle, "An algorithm for total variation minimization and applications," *Journal of Mathematical imaging and vision*, vol. 20, pp. 89–97, 2004.
60. A. Abraham, F. Pedregosa, M. Eickenberg, P. Gervais, A. Mueller, J. Kossaifi, A. Gramfort, B. Thirion, and G. Varoquaux, "Machine learning for neuroimaging with scikit-learn," *Frontiers in neuroinformatics*, vol. 8, p. 14, 2014.
61. D. J. Blackburn, Y. Zhao, M. De Marco, S. M. Bell, F. He, H.-L. Wei, S. Lawrence, Z. C. Unwin, M. Blyth, J. Angel *et al.*, "A pilot study investigating a novel non-linear measure of eyes open versus eyes closed EEG synchronization in people with Alzheimer's disease and healthy controls," *Brain sciences*, vol. 8, no. 7, p. 134, 2018.
62. L. Rui, H. Nejati, and N.-M. Cheung, "Dimensionality reduction of brain imaging data using graph signal processing," *2016 IEEE International Conference on Image Processing (ICIP)*. IEEE, 2016, pp. 1329–1333.
63. G. Fracastoro, D. Thanou, and P. Frossard, "Graph transform optimization with application to image compression," *IEEE Transactions on Image Processing*, vol. 29, pp. 419–432, 2019.
64. A. Anis, A. Gadde, and A. Ortega, "Towards a sampling theorem for signals on arbitrary graphs," *2014 IEEE International Conference on Acoustics, Speech and Signal Processing (ICASSP)*. IEEE, 2014, pp. 3864–3868.
65. G. Puy, N. Tremblay, R. Gribonval, and P. Vandergheynst, "Random sampling of bandlimited signals on graphs," *Applied and Computational Harmonic Analysis*, vol. 44, no. 2, pp. 446–475, 2018.
66. J. S. Walker, "Fourier analysis and wavelet analysis," *Notices of the AMS*, vol. 44, no. 6, pp. 658–670, 1997.
67. M. G. Preti and D. Van De Ville, "Decoupling of brain function from structure reveals regional behavioral specialization in humans," *Nature communications*, vol. 10, no. 1, p. 4747, 2019.
68. A. Pentari, G. Tzagkarakis, K. Marias, and P. Tsakalides, "Graph denoising of impulsive EEG signals and the effect of their graph representation," *Biomedical Signal Processing and Control*, vol. 78, p. 103886, 2022.
69. J. Zhou, G. Cui, S. Hu, Z. Zhang, C. Yang, Z. Liu, L. Wang, C. Li, and M. Sun, "Graph neural networks: A review of methods and applications," *AI open*, vol. 1, pp. 57–81, 2020.
70. F. M. Bianchi, D. Grattarola, L. Livi, and C. Alippi, "Graph neural networks with convolutional ARMA filters," *IEEE transactions on pattern analysis and machine intelligence*, vol. 44, no. 7, pp. 3496–3507, 2021.
71. T. Song, W. Zheng, P. Song, and Z. Cui, "EEG emotion recognition using dynamical graph convolutional neural networks," *IEEE Transactions on Affective Computing*, vol. 11, no. 3, pp. 532–541, 2018.

72. P. Zhong, D. Wang, and C. Miao, "EEG-based emotion recognition using regularized graph neural networks," *IEEE Transactions on Affective Computing*, vol. 13, no. 3, pp. 1290–1301, 2020.
73. A. Demir, T. Koike-Akino, Y. Wang, M. Haruna, and D. Erdogmus, "EEG-GNN: Graph neural networks for classification of electroencephalogram (EEG) signals," *2021 43rd Annual International Conference of the IEEE Engineering in Medicine & Biology Society (EMBC)*. IEEE, 2021, pp. 1061–1067.
74. S. Tang, J. A. Dunnmon, K. Saab, X. Zhang, Q. Huang, F. Dubost, D. L. Rubin, and C. Lee-Messer, "Self-supervised graph neural networks for improved electroencephalographic seizure analysis," *arXiv preprint arXiv:2104.08336*, 2021.
75. G. Kalantar and A. Mohammadi, "Graph-based dimensionality reduction of EEG signals via functional clustering and total variation measure for BCI systems," *2018 40th Annual International Conference of the IEEE Engineering in Medicine and Biology Society (EMBC)*. IEEE, 2018, pp. 4603–4606.
76. G. Mateos, S. Segarra, A. G. Marques, and A. Ribeiro, "Connecting the dots: Identifying network structure via graph signal processing," *IEEE Signal Processing Magazine*, vol. 36, no. 3, pp. 16–43, 2019.
77. X. Dong, D. Thanou, M. Rabbat, and P. Frossard, "Learning graphs from data: A signal representation perspective," *IEEE Signal Processing Magazine*, vol. 36, no. 3, pp. 44–63, 2019.
78. V. Kalofolias, "How to learn a graph from smooth signals," *Artificial Intelligence and Statistics*. PMLR, 2016, pp. 920–929.
79. P. Mathur and V. K. Chakka, "Graph signal processing based cross-subject mental task classification using multi-channel EEG signals," *IEEE Sensors Journal*, vol. 22, no. 8, pp. 7971–7978, 2022.
80. A. Einizade, M. Mozafari, M. Rezaei-Dastjerdehei, E. Aghdaei, A. M. Mijani, and S. H. Sardouie, "Detecting ADHD children based on EEG signals using graph signal processing techniques," *2020 27th National and 5th International Iranian Conference on Biomedical Engineering (ICBME)*. IEEE, 2020, pp. 264–270.
81. J. Jeong, "EEG dynamics in patients with Alzheimer's disease," *Clinical neurophysiology*, vol. 115, no. 7, pp. 1490–1505, 2004.
82. S. P. van den Broek, F. Reinders, M. Donderwinkel, and M. Peters, "Volume conduction effects in EEG and MEG," *Electroencephalography and clinical neurophysiology*, vol. 106, no. 6, pp. 522–534, 1998.
83. C. M. Michel, M. M. Murray, G. Lantz, S. Gonzalez, L. Spinelli, and R. G. De Peralta, "EEG source imaging," *Clinical neurophysiology*, vol. 115, no. 10, pp. 2195–2222, 2004.
84. J.-M. Schoffelen and J. Gross, "Source connectivity analysis with MEG and EEG," *Human brain mapping*, vol. 30, no. 6, pp. 1857–1865, 2009.
85. M. Lai, M. Demuru, A. Hillebrand, and M. Fraschini, "A comparison between scalp-and source-reconstructed EEG networks," *Scientific reports*, vol. 8, no. 1, p. 12269, 2018.
86. A. Hagberg, P. Swart, and D. S Chult, "Exploring network structure, dynamics, and function using networkx," Los Alamos National Lab.(LANL), Los Alamos, NM (United States), Tech. Rep., 2008.
87. R. W. Thatcher, "Coherence, phase differences, phase shift, and phase lock in EEG/ERP analyses," *Developmental neuropsychology*, vol. 37, no. 6, pp. 476–496, 2012.

Electroencephalographic Changes in Carotid Endarterectomy Correlated with Ischemia

Shyam Visweswaran, Katherine C. Du, Vinay Pedapati, Amir I. Mina, Allison M. Bradley, Jessi U. Espino, Kayhan Batmanghelich, and Parthasarathy D. Thirumala

1 Introduction

1.1 Intraoperative Ischemia and Stroke

Despite advances in medical optimization and surgical techniques, the risk of perioperative stroke, which occurs during or within 30 days of surgery, has risen over the past decade [1]. Every year in the USA, approximately 50,000 patients suffer from perioperative stroke, most of which occur during or just after surgery [2]. Significant disability or death from a perioperative stroke imposes a significant financial and economic burden on individuals and society [3, 4]. Perioperative stroke can occur intraoperatively (during surgery) or postoperatively (after surgery). The incidence of stroke during surgery is as high as 6% for cardiovascular and neurological surgeries and as low as 1% for general surgeries [2]. Intraoperative stroke generally occurs from cerebral ischemia resulting from decreased blood flow to the brain [5]. Sometimes, decreased blood flow may be due to a blood clot traveling to the brain. At other times, decreased blood flow can occur due to a decrease in blood pressure, resulting in hypoperfusion of the brain. Therefore,

S. Visweswaran (✉) · K. C. Du · V. Pedapati · A. I. Mina · A. M. Bradley · J. U. Espino
Department of Biomedical Informatics, University of Pittsburgh, Pittsburgh, PA, USA
e-mail: shv3@pitt.edu; katherine.du@pitt.edu; mina.amir@medstudent.pitt.edu; ambradley@pitt.edu; jue@pitt.edu

K. Batmanghelich
Department of Electrical and Computer Engineering, Boston University, Boston, MA, USA
e-mail: batman@bu.edu

P. D. Thirumala
Department of Neurological Surgery, University of Pittsburgh, Pittsburgh, PA, USA
e-mail: thirumalpd@upmc.edu

A. Ahmed, J. Picone (eds.), *Machine Learning Applications in Medicine and Biology*,
https://doi.org/10.1007/978-3-031-51893-5_2

intraoperative monitoring for cerebral ischemia is essential to enhance surgical safety and reduce the incidence of strokes.

1.2 Cerebral Blood Flow and Ischemia

During surgery, uninterrupted cerebral blood flow (CBF) is critical for brain function since it ensures adequate delivery of oxygen and nutrients. Because of the limited capacity of the brain to store oxygen and nutrients, disruptions to CBF, even for a few minutes, can cause debilitating brain damage [6]. Initially, when CBF is reduced or stopped, it results in reversible brain damage from ischemia [7]. If CBF is not restored rapidly, the brain damage becomes irreversible due to the death of brain cells and results in a stroke [8, 9]. Therefore, taking preventive measures as soon as possible while the patient is experiencing cerebral ischemia will reduce the risk of stroke and irreversible brain damage.

1.3 Electroencephalography and Ischemia

Brain function can be continuously and noninvasively measured using electroencephalography (EEG). Reduced CBF causes changes in the metabolic and electrical activity of the brain, which are reflected in the EEG [8]. When CBF decreases to 25–35 ml/100 g/min, EEG activity of higher frequencies (8–12 Hertz (Hz)) declines, and when CBF drops further to 18–25 ml/100 g/min, EEG activity of lower frequencies (4–8 Hz) increases (Table 1). Further decreases in CBF to 12–18 ml/100 g/min result in increased activity of yet lower frequencies (1–4 Hz). When CBF falls below 12–18 ml/100 g/min, cerebral electrical activity suffers, which can be reversed by increasing CBF [10]. This corresponds to ischemia, resulting in reversible brain damage. However, prolonged levels of CBF at this level and further decreases in CBF to less than 10–12 ml/100 g/min cause the EEG to go silent, which cannot be reversed by increasing CBF. This corresponds to a stroke resulting in irreversible brain damage. Depending on the severity of the CBF reduction, the transition from ischemia to stroke may occur quickly in minutes or is prolonged for hours.

1.4 Intraoperative Neurophysiological Monitoring with EEG

Intraoperative neurophysiological monitoring (IONM) with EEG is routinely employed to detect ischemia during high-risk neurological and cardiac surgeries [11, 12]. IONM with EEG serves a similar purpose as routine cardiac monitoring with electrocardiography (EKG) for cardiac ischemia. However, fewer than 10%

Table 1 Electroencephalographic (EEG) changes related to cerebral blood flow (CBF)

Cerebral blood flow (CBF)	EEG change
35–50 ml/100 g/min	Normal
25–35 ml/100 g/min	Loss of higher frequency activity (8–12 Hz)
18–25 ml/100 g/min	Increase in lower frequency activity (4–8 Hz)
Ischemia threshold	
12–18 ml/100 g/min	Further increase in lower frequency activity (0–4 Hz)
Stroke threshold	
<10–12 ml/100 g/min	Electrical silence (silent EEG)

of cardiovascular and neurological surgeries are monitored for brain function with IONM. One reason for the lack of utilization of IONM is the need for real-time manual visual review of EEG by a skilled neurophysiologist. This means that the neurophysiologist needs to be available in real time to read the EEG during all cardiovascular and neurological surgeries. Another reason for the limited use is the lack of standard, routine screening protocols for patients presenting a risk of intraoperative stroke; such screening would identify the best candidates for IONM. Further, continuous visual monitoring is mentally demanding and tiring and can be error prone. The lack of IONM translates to delays in detecting ischemia and stroke and in obtaining brain imaging and starting treatment. In contrast, cardiac function is monitored with EKG in all surgeries, and high-risk patients are identified and managed with standard protocols [13]. Although all patients undergoing surgery have their hearts monitored with EKG, the brain is not routinely monitored in most surgeries, except for a few procedures such as carotid endarterectomy. Thus, there is a great need for automated methods of continuous real-time monitoring for cerebral ischemia in other high-risk surgeries beyond carotid endarterectomy.

1.5 Carotid Artery Endarterectomy

Carotid artery endarterectomy (CEA) is a surgical procedure that removes plaque from inside the common or internal carotid artery [14]. Plaque in the artery narrows the artery, increasing the risk of stroke, and removal of the plaque restores normal blood flow to the brain. During CEA, the afflicted artery is temporarily clamped to allow the plaque to be removed. When an artery is clamped, blood flow through the affected vessel is temporarily stopped. Other blood vessels continue to supply blood to the brain during this time, but this may not be enough. There is a risk of ischemic stroke during CEA if blood flow drops too far. Because EEG changes are sufficiently sensitive and specific to detect changes early on, intraoperative EEG is routinely used in CEA to monitor for cerebral ischemia, allowing the surgeon to take action to restore CBF to safe levels [15, 16].

If ischemia occurs during surgery, the EEG of the patient will change accordingly (Table 1). When changes are detected, the monitoring neurophysiologist relays the information to the surgeon, who then decides how to proceed. Typically, the surgeon will insert a shunt to redirect blood flow around the artery being operated on, ensuring that the brain receives adequate blood. Once the shunt is in place, the EEG usually returns to normal within a few minutes, allowing the surgeon to proceed. Thus, the use of IONM during CEA enables the surgeon to selectively use shunting in only those patients in whom clamping causes significant cerebral ischemia; selective shunting is preferred to shunting all patients due to the increased risk of surgical complications associated with shunting [17].

2 Related Works

2.1 EEG and Continuous EEG

The EEG measures and records the brain's electrical activity in terms of voltage over time, and the recording is obtained by attaching electrodes to the scalp in standard locations [18]. IONM typically employs eight electrodes attached to four locations on the left side and four locations on the right side of the head. Each electrode location is labeled with a letter and a number. The letter indicates the region of the brain beneath the electrode, such as F for the frontal lobe, P for the parietal lobe, T for the temporal lobe, and O for the occipital lobe. Even numbers denote the right side of the head, while odd numbers denote the left side. Thus, F3 is an electrode location above the left frontal lobe of the brain, whereas O2 is an electrode location above the right occipital lobe of the brain. Typically, voltage is measured across two electrodes and displayed as a channel. Thus, the F3-P3 channel measures the voltage across the electrodes F3 and P3 (Fig. 1). In typical IONM, eight distinct channels are displayed, with each channel displaying the voltage difference between two electrodes comprising that channel (Fig. 1).

Technological advances in the 1990s enabled the EEG to be digitized and recorded in electronic data storage, allowing for continuous EEG (cEEG) recordings over hours to days [19]. Unlike routine EEG, which records brain activity for only 20–30 minutes, cEEG provides a dynamic view of brain activity over extended time periods, typically ranging from hours to days, and was originally used to assess brain function in critically ill and comatose patients. Since the late 1990s, cEEG has been increasingly used in intensive care units to detect and monitor seizures, brain injury, intracranial bleeding, brain infection, and cerebral ischemia [20, 21]. Another application for cEEG is in the operating room, where anesthesiologists use it to monitor the depth of anesthesia, and neurophysiologists use it to monitor cerebral ischemia [22].

As a signal that varies in time, the EEG can be viewed, analyzed, and interpreted in two distinct domains: the time domain and the frequency domain. The EEG signal is commonly displayed in the time domain, with time, typically measured in seconds (s), plotted on the x-axis and amplitude, typically measured in microvolts (μV)

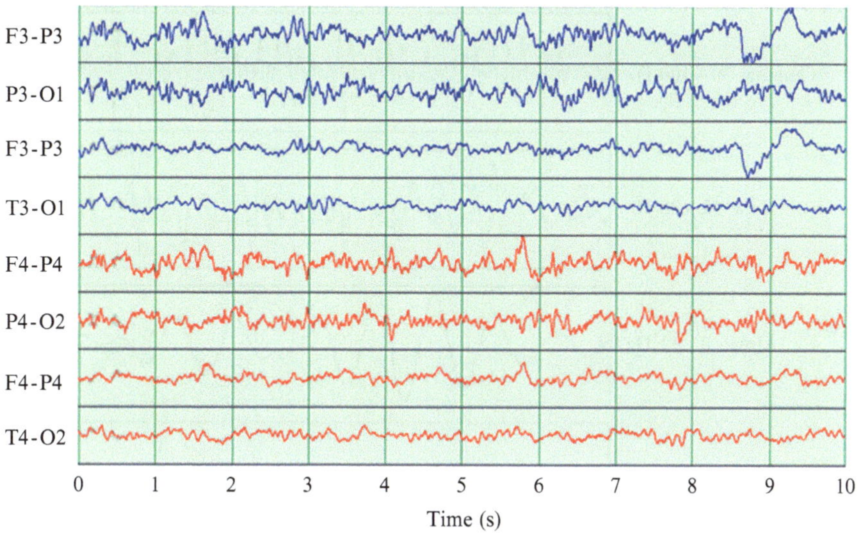

Fig. 1 An eight-channel EEG segment is shown with blue-colored signals for the left side of the brain and red-colored signals for the right side. Channels F3-P3, P3-O1, F3-T3, and T3-O1 are for the left side of the brain, while channels F4-P4, P4-O2, F4-T4, and T4-O2 are for the right side

plotted on the *y*-axis, and a typical cEEG display shows recordings of 10–20 seconds at a time. Visual interpretation focuses on amplitude and morphology changes in EEG waveforms over time in specific frequency bands [7, 12]; the clinically used frequency bands in increasing order of frequency are delta (0–4 Hz), theta (4–8 Hz), alpha (8–12 Hz), beta (12–30 Hz), and gamma (30–100 Hz) (Fig. 2). As frequency increases, the amplitude of the EEG signal typically decreases. Thus, the delta and theta bands have lower frequencies but greater amplitudes (and are referred to by neurophysiologists as slow activity), whereas the alpha, beta, and gamma bands have higher frequencies but smaller amplitudes (and are referred to as fast activity). Typically, higher levels of arousal are associated with higher frequencies. The gamma band has the highest frequency and lowest amplitude and is associated with enhanced perception, such as when a person is concentrating intensely on a task. The beta band is related to normal awake consciousness, such as when a person is actively conversing. The alpha band is associated with relaxed consciousness, such as when an individual is reflective or introspective. The theta band is associated with reduced consciousness, such as when a person is daydreaming. The delta band has the lowest frequency and greatest amplitude and is associated with deep sleep.

To read the amplitudes of the frequencies, the complementary frequency domain view is used. In this view, frequency, typically measured in Hz, is plotted on the *x*-axis, and amplitude, typically measured in μV, is plotted on the *y*-axis. However, it is more common to display the power, which is the square of the amplitude, typically measured in μV [2], on the *y*-axis. This is called the EEG power spectrum and is computed by applying a fast Fourier transform (FFT) to the EEG signal. When

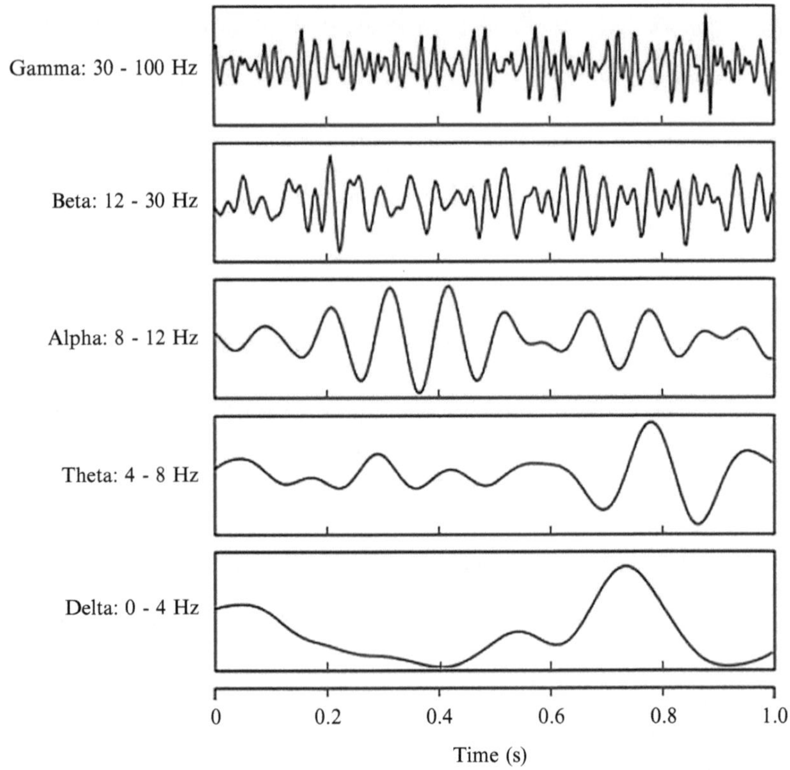

Fig. 2 Illustration of the clinically used EEG frequency bands: delta, theta, alpha, beta, and gamma

applied to a segment of an EEG recording, FFT decomposes the segment into its component frequencies and the corresponding power in each frequency. As shown in Fig. 3, the power spectrum directly reveals the power of each frequency, which cannot be seen easily in the time domain [23].

2.2 Spectral Displays of cEEG

The power spectrum is limited since it displays the power levels by frequency for only a single segment of the EEG recording. However, the use of the spectrum in IONM requires its computation on successive EGG segments; this results in the generation of a spectrogram [24, 25]. Because the neurophysiologist must be able to view successive past spectra simultaneously, methods for displaying the spectrogram in a time-constrained manner have been developed. The two most widely used time-compressed displays are the compressed spectral array (CSA) and

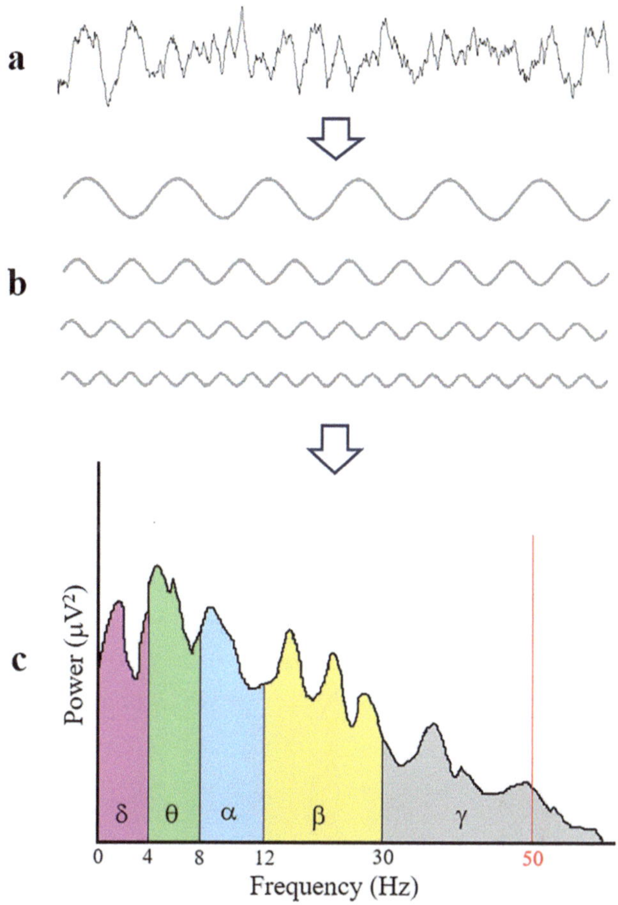

Fig. 3 A diagrammatic representation of the creation of a power spectrum. FFT is used to decompose an EEG segment (**a**) into sine waves of varying frequencies and amplitudes (**b**). The amplitude is converted to power and then plotted against frequency to generate a power spectrum (**c**). The area of the curve at specific frequency intervals can be used to calculate the power of each frequency band (δ = delta, θ = theta, α = alpha, β = beta, and γ = gamma). The red vertical line at 50 Hz denotes the spectral edge frequency 90 (SEF90)

the density spectral array (DSA) [25, 26]. The CSA displays the frequency, power, and time data array as a pseudo-three-dimensional topographic plot. In the CSA plot shown in Fig. 4, the *x*-axis represents the frequency in Hz, the *y*-axis denotes time in hour:minute:second format, and power is displayed as a line with the most recent segment shown in front of the previous segment and shifted down the time axis. The DSA presents the same data as a two-dimensional colored heatmap. In the DSA plot shown in Fig. 5, the *x*-axis represents the frequency in Hz, the *y*-axis denotes time in hour:minute:second format, and power is represented by color-coded bands, with

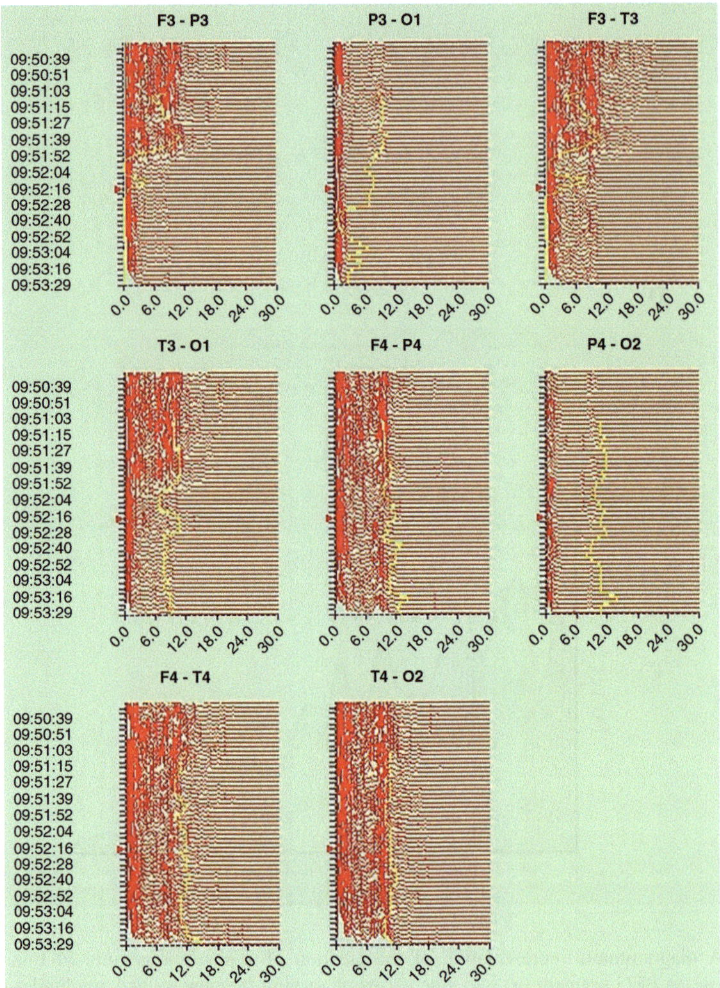

Fig. 4 Eight compressed spectral array (CSA) plots corresponding to an eight-channel EEG. The x-axis represents the frequency in Hz, the y-axis denotes time in hour:minute:second format, and power is displayed as a line with the most recent segment shown in front of the previous segment and shifted down the time axis

blue representing low power and red representing high power. Similar to the CSA plot, the time axis of the DSA plot displays the progression of the power spectra of successive segments [27]. Although both plots express the same information, the DSA is more compact, while the CSA permits better resolution of the power levels. The "mountain-range" aspect of the CSA plot aids the neurophysiologist in quickly identifying the EEG frequencies of the highest powers, whereas the colored bands of the DSA plot aid the neurophysiologist in monitoring changes in power.

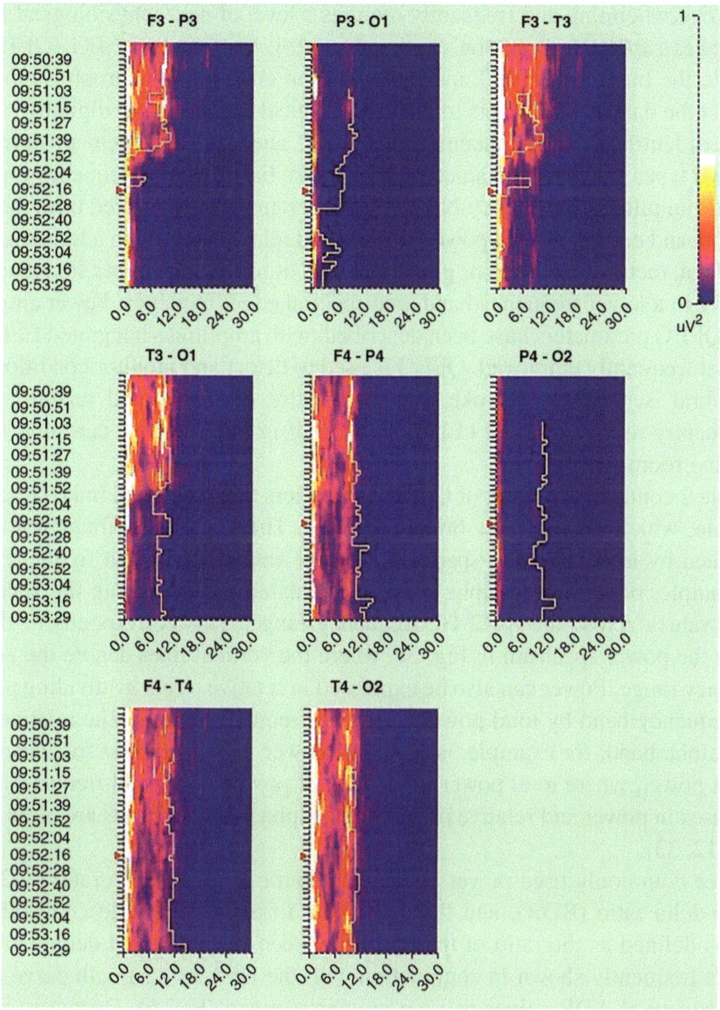

Fig. 5 Eight density spectral array (DSA) plots corresponding to an eight-channel EEG. The *x*-axis represents the frequency in Hz, the *y*-axis denotes time in hour:minute:second format, and power is represented by color-coded bands, with blue representing low power and red representing high power

2.3 *Quantitative EEG*

The use of cEEG, such as during intraoperative monitoring, generates substantial volume of data. Typically, the EEG display shows 10–20 seconds of data at a time, and 1 hour of monitoring produces 180–360 displays of data. Although time-compressed displays of plots in the frequency domain aid in understanding the evolution of power changes, continuous monitoring of such a large volume of data can be overwhelming and frequently requires a level of neurophysiological training that makes rapid interpretation challenging [28]. Quantitative EEG (QEEG) can alleviate the burden of cEEG analysis and even enable non-neurophysiologists to interpret the data. QEEG refers to the mathematical processing of digitally recorded EEG to identify patterns, quantify variations, and compare them to norms [29, 30]. Two types of QEEG parameters exist: those based on the frequency and those based on amplitude. Frequency-based QEEG parameters are derived from the power spectrum and consist of total power, powers and relative powers in a frequency band (e.g., delta, theta, alpha, beta, or gamma), ratios of power in a higher frequency band to power in a lower frequency band, and spectral edge frequency. Fewer amplitude-based QEEG parameters have been described, with amplitude-integrated EEG being the most frequently employed. QEEG is used to detect and monitor conditions such as cerebral ischemia and stroke, encephalopathy, dementia, and seizures [31]. It is frequently used as part of cEEG monitoring in the intensive care unit and the operating room [32].

Table 2 contains examples of QEEG parameters that have been linked to cerebral ischemia, which we will now briefly describe. The power of a frequency band is calculated by integrating the spectrum's power values across that frequency band, for example, power in the alpha band is calculated by integrating the spectrum's power values across the 8–12 Hz frequency range. This corresponds to the blue area in the power spectrum in Fig. 3c, where the vertical lines denote the 8–12 Hz frequency range. Power can also be expressed in relative terms by dividing power in one frequency band by total power across all frequency bands. The relative power in the alpha band, for example, is a ratio of power in the 8–12 Hz frequency range to total power, where total power is the sum of powers across all frequency bands. Decreases in power and relative power in the alpha and beta bands are often seen in CEA [12, 33].

Three commonly used power ratios include the alpha-to-delta ratio (ADR), the beta-to-delta ratio (BDR), and the alpha-beta to delta-theta ratio (ABDTR). The ADR is defined as the ratio of the power between the alpha and delta bands. The ADR is frequently shown in conjunction with the DSA as a smooth curve derived from sequential ADR values using a moving average (Fig. 6). Decreases in ADR values are associated with cerebral ischemia that may occur due to attenuation of blood flow to the brain during surgery [34] and in subarachnoid hemorrhage, which is bleeding in the space between the brain and the surrounding membrane [35, 36]. The BDR is defined as the ratio of the power between the beta and delta bands. Decreases in BDR are associated with intraoperative cerebral ischemia [34]. The ABDTR is the sum of alpha and beta powers divided by the sum of delta and theta powers. Decreases in ABDTR values have been linked to large strokes, and larger decreases have been associated with mortality following a large stroke [37]. The spectral edge frequency (SEF) is derived from the power spectrum and is the frequency below which a specified proportion (often 90%) of the power in the EEG signal is contained. In Fig. 3c, the red vertical line at a frequency of 50 Hz denotes SEF90 since 90% of the colored area lies to the left of the line. SEF has been found to be useful in monitoring the depth of anesthesia [38].

Table 2 Selected examples of QEEG parameters with brief descriptions

Parameter type	QEEG parameter	Abbreviation	Description
Total power			The sum of power across all frequencies in an EEG segment. A decrease in total power indicates a decrease in amplitude
Power	Delta power Theta power Alpha power Beta power Gamma power		The power or amount of activity in the specified frequency band in an EEG segment
Relative power	Relative delta power Relative theta power Relative alpha power Relative beta power Relative gamma power		The ratio of the power in the specified frequency band in an EEG segment to the total power
Power ratio	Alpha-to-delta ratio Beta-to-delta ratio Alpha-beta to delta-theta ratio	ADR BDR ABDTR	The ratio of powers in two specified frequency bands in an EEG segment; often, the numerator is the power of a higher frequency band, and the denominator is the power of a lower frequency band
Spectral edge frequency	Spectral edge frequency 90	SEF90	Spectral edge frequency is the highest frequency below which a specified amount of power is present in an EEG segment. SEF90 is the frequency below which 90% of the power is present
Amplitude-integrated EEG		aEEG	The range of the amplitude (difference between highest and lowest voltages) in an EEG segment

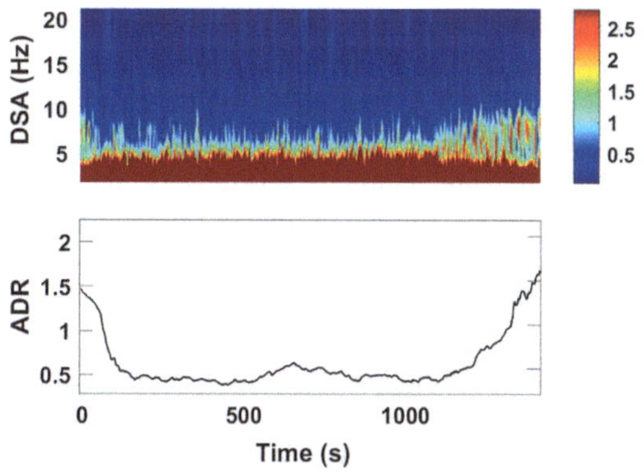

Fig. 6 Density spectral array (DSA) and corresponding alpha to delta ratio (ADR) of a single channel of EEG recording

Amplitude-integrated EEG (aEEG) is a parameter based on the EEG's amplitude as opposed to its frequency. The EEG signal is processed to extract the minimum and maximum amplitude of brief segments of 1–2 seconds in length, and the results are displayed in a time-compressed plot. Prior studies have shown that aEEG is useful in detecting and monitoring seizures [39].

2.4 EEG Changes Related to Cerebral Ischemia

Foreman et al. reported in a review that cEEG monitoring for detecting and monitoring cerebral ischemia and stroke is most commonly used in the intensive care unit and the operating room [35]. In the intensive care unit, strokes and delayed cerebral ischemia in subarachnoid bleeding can be associated with changes in specific QEEG parameters. In a stroke, QEEG parameters that are decreased include the relative delta power, relative alpha power, ADR, and ABDTR values. Using a database containing EEG recordings of 31 patients with stroke and 10 control subjects, Sheorajpanday et al. measured a range of QEEG parameters and found that decreases in relative delta power, ADR, and ABDTR values were most strongly correlated with stroke [40]. In delayed cerebral ischemia, Classen et al. reported that the decrease in ADR values from baseline was the strongest predictor, followed by relative alpha power [41].

In the operating room, monitoring for ischemia is performed in high-risk cardiovascular and neurological surgeries, and in particular routine cEEG monitoring is performed in CEA. Visual analysis of the cEEG, CSA, and DSA is used to identify patterns associated with cerebral ischemia. Visual changes in the EEG are strongly related to changes in the CBF, as described in Sect. 1.3; as CBF decreases, activity of higher frequencies progressively decrease, and as CBF decreases further, activity of lower frequencies gradually increase. In the context of CEA, EEG changes frequently occur within 20 seconds of the artery being clamped, with the majority occurring within 2 minutes, in patients whose CBF is critically dependent on the clamped artery. In 3–5% of CEA patients, the CBF drops instantly to the ischemic threshold accompanied by corresponding changes in the EEG; in these patients, a temporary bypass shunt is used to continue the surgery [42]. Visser et al. analyzed data from 92 CEA patients who underwent transcranial Doppler ultrasonography to measure blood flow velocity in intracranial arteries in addition to cEEG monitoring [43]. Following clamping, severe ischemia was associated with a decrease in alpha and beta band activity followed by an increase in delta and theta band activity, whereas mild ischemia was associated with only a decrease in alpha and beta band activity. Some patients demonstrated an increase in fast activity (alpha and beta) and a decrease in slow activity (delta and theta), which was attributed to arousal caused by pain and hemodynamic stress induced by clamping. Kamitaki et al. analyzed cEEG recordings of 118 CEA patients and found that decreases in alpha, beta, and theta power of 52.1%, 41.6%, and 36.4%, respectively, were the strongest predictors of ischemia following clamping [34]. Furthermore, De Almeida et al. describe the often-used "10–20" clinical rule, which states that a significant EEG change of up to 10 minutes can be tolerated with minimal neurological deficits; that after 10 minutes, the incidence of neurological deficits increases; and that by 20 minutes, neurological deficits are almost always present [44].

Performing visual analysis for extended periods of time in real time is difficult. Furthermore, such analysis requires practice, and subtle changes in patterns can be missed with only visual inspection. Consequently, multiple QEEG parameters for detecting ischemia during CEA and other surgeries have been evaluated. Kamitaki et al. analyzed cEEG recordings of 118 CEA patients and found that ADR, BDR, and aEEG values decreased during ischemia following clamping [34]. Using a database of 149 CEA patients, Laman et al. showed that SEF 90% was the best single QEEG parameter among 12 parameters (powers, relative powers, and spectral edge frequencies) for discriminating between patients who required a shunt and those who did not [45]. According to Foreman et al., the utility of QEEG parameters in CEA depends on the type of anesthesia administered [35]. For instance, changes in SEF are useful for detecting ischemia under isoflurane anesthesia, whereas changes in the relative delta power are accurate in patients receiving propofol.

3 Methods

3.1 Data and Preprocessing

We acquired a database of intraoperative cEEG recordings of 1551 patients who underwent CEA in UPMC hospitals in Pittsburgh between the years 2009 and 2017. All cEEG recordings included the four channels F3-P3, P3-O1, F3-T3, and T3-O1 on the left side and the four channels F4-P4, P4-O2, F4-T4, and T4-O2 on the right side. A neurophysiologist reviewed the recordings, and a patient was labeled as ischemic-change if there were visual changes indicative of ischemia in the 10-minute post-clamp recording; otherwise, the patient was labeled as no-ischemic-change.

We extracted a 10-minute segment post-clamp from each recording; this segment was divided into 20-second intervals from which ten QEEG parameters were derived separately for the ipsilateral (side of clamp) and contralateral (opposite side) sides. We obtained QEEG z-scores for each channel from a 20-second interval normalized using a 120-second pre-clamp baseline EEG for a patient. We identified a 120-second pre-clamp baseline EEG with minimal noise and variance for each patient in order to estimate the baseline distribution parameters. We derived ten averaged QEEG parameters for statistical analyses, including alpha, beta, theta, and delta power values; ADR, BDR, and ABDTR power ratio values; SEF90; and aEEG values obtained by averaging the z-scores of four channels from one side of the head. For instance, the alpha QEEG parameters were obtained by averaging the

z-scores of the left-sided channels F3-P3, P3-O1, F3-T3, and T3-O1 and the right-sided channels F4-P4, P4-O2, F4-T4, and T4-O2, respectively.

3.2 Analysis of Correlation of Quantitative EEG (QEEG) with Ischemia

We analyzed post-clamp QEEG changes that correlate with ischemia, for the entire database as well as ischemia-related QEEG changes in the sexes separately.

We divided the database of 1551 patients into 84 ischemic-change patients and 1469 patients with no-ischemic-change. Since each 10-minute segment was partitioned into thirty 20-second intervals, we calculated 30 z-scores from the spectrogram from the 20-second intervals for each QEEG parameter. From this series of 30 values, we identified the maximum and minimum z-scores for each QEEG parameter as the peak and trough values, respectively. We used the nonparametric Mann-Whitney U test to identify QEEG parameters that were statistically significantly different at the 0.05 level between the ischemic-change and no-ischemic-change patient groups; we used a nonparametric test because the QEEG parameters were not normally distributed. Though the database is skewed due to the unequal sample sizes of the ischemic-change and no-ischemic-change groups, the variances in the two groups were comparable. We conducted separate statistical analyses on the QEEG parameters for the ipsilateral and contralateral sides of the head.

We performed similar statistical analyses between the ischemic-change and no-ischemic-change patient groups for each sex. From the database of 1551 patients, we discarded 39 patients whose sex was not recorded, and we divided the remaining 1512 patients into 641 female and 903 male patients. For the sex-stratified analyses, we restricted the analyses to the ipsilateral QEEG parameters.

3.3 Analysis of the Effect of Age, Sex, and Side of Surgery on Ischemia

From the database of 1551 patients, we retained 1512 patients for whom age, sex, and side of surgery were available. We performed logistic regression analysis with age, sex, and side of surgery as independent variables and ischemic change as the dependent variable. The analysis was conducted using age (range from 41 to 96 years old), sex (630 females and 882 males), and side of surgery (746 with left-sided CEA and 766 patients with right-sided CEA), with ischemic-change as the dependent variable (82 with ischemic-change and 1430 with no-ischemic-change). For the independent variables, age was input as a whole number, sex was specified as male = 0 and female = 1, and side of surgery was specified as left-sided CEA = 0 and right-sided CEA = 1. The dependent variable was specified as no-ischemic-change = 0 and ischemic-change = 1 based on the neurophysiologist's labeling of the 10-minute post-clamp recording. We fitted maximum-likelihood dichotomous logistic models to report odds ratios and the 95% confidence intervals for each independent variable.

4 Results and Discussion

4.1 Correlation of Quantitative EEG with Ischemia

In Fig. 7, the median *z*-scores for ten ipsilateral QEEG parameters are compared between the patient groups with and without ischemic change. Ipsilateral QEEG trough values were significantly decreased from pre-clamp baseline for all ten parameters in the ischemic-change group when compared with the no-ischemic-change group. Ipsilateral peak values were also significantly decreased for nine of the ten parameters, excluding the delta parameter, in the ischemic-change group

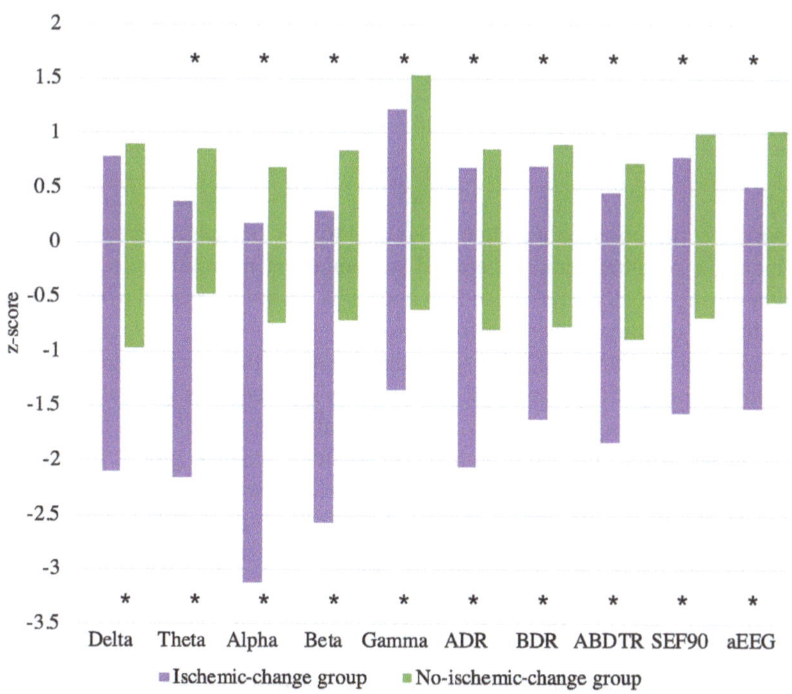

Fig. 7 Comparison of median *z*-scores for ipsilateral QEEG parameters between ischemic-change and no-ischemic-change groups. Positive *z*-scores denote peak values, whereas negative *z*-scores denote trough values

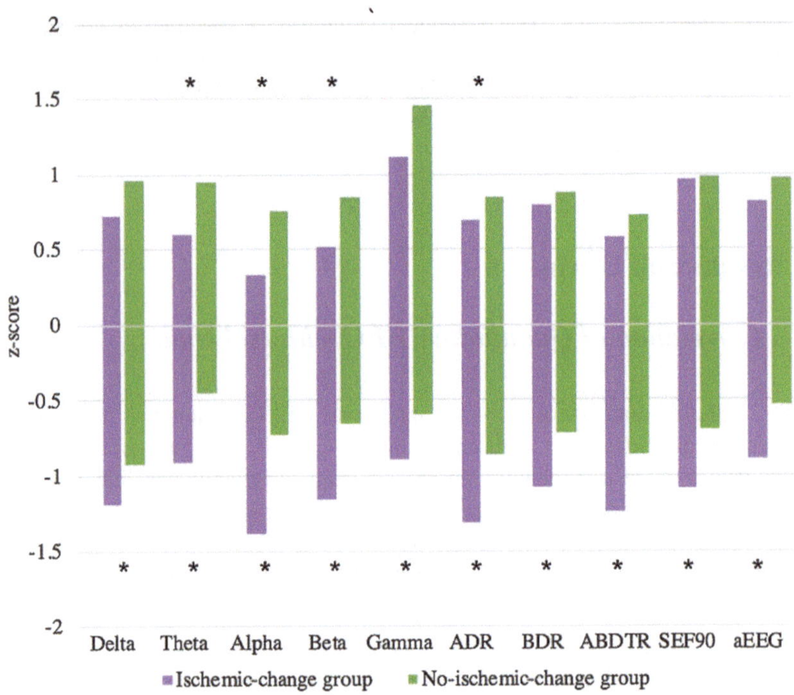

Fig. 8 Comparison of median z-scores for contralateral QEEG parameters between ischemic-change and no-ischemic-change groups. Positive z-scores denote peak values, whereas negative z-scores denote trough values

compared to the no-ischemic-change group [46]. A decreased trough value is a key indicator of intraoperative cerebral ischemia that neurophysiologists visually monitor during surgery. These results support this observation as all troughs showed a statistically significant decrease from baseline.

Figure 8 is similar to Fig. 7 except that it shows the median z-scores for the ten contralateral QEEG parameters. Like ipsilateral parameters, contralateral QEEG trough values were significantly decreased for all ten parameters in the ischemic-change group when compared with the no-ischemic-change group. Compared to the ipsilateral peak values, fewer contralateral QEEG parameters were significantly decreased; the parameters that were significantly decreased included theta, alpha, beta, and ADR values [46]. Our results corroborate prior studies that a decrease in QEEG parameters is more common than an increase after carotid clamping [33, 34]. Kamitaki et al. reported ipsilateral decreases in eight QEEG parameters, including alpha, beta, theta, delta power, ADR, BDR, aEEG, and SEF90 values [34]. Our results corroborate these findings and extend them to two additional QEEG parameters, namely, gamma power and ABDTR values.

Figures 9 and 10 compare the ten ipsilateral QEEG parameters between the patient groups with and without ischemic change for females and males, respec-

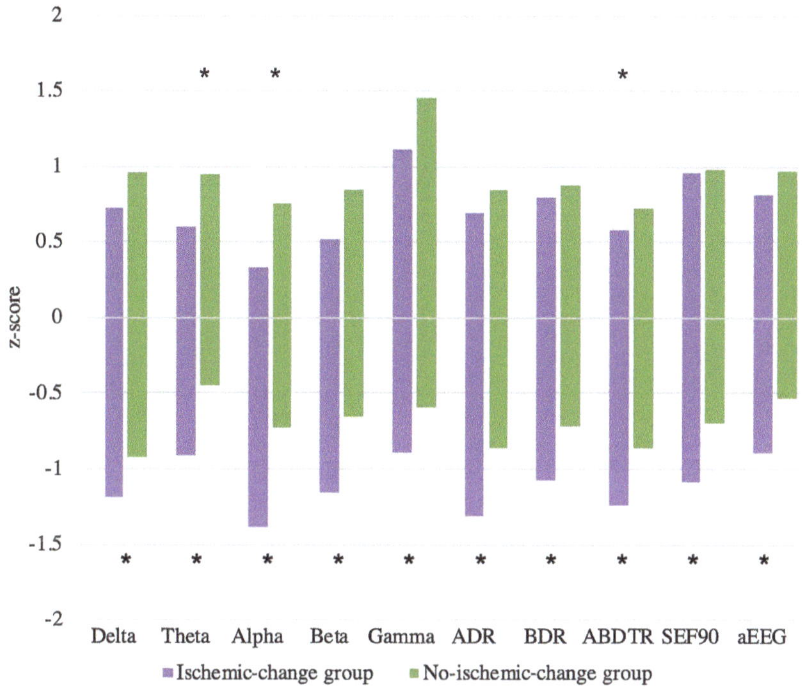

Fig. 9 Comparison of median z-scores for ipsilateral QEEG parameters between ischemic-change and no-ischemic-change groups in females. Positive z-scores denote peak values, whereas negative z-scores denote trough values

tively. In both males and females, ipsilateral QEEG trough values were significantly decreased from the pre-clamp baseline for all ten parameters in the ischemic-change group when compared with the no-ischemic-change group. In contrast to the trough values, peak values were significantly decreased for only three QEEG parameters in females (theta, alpha, and ABDTR) and four QEEG parameters in males (theta, alpha, beta, and gamma) [47]. The sex-stratified results show that the QEEG troughs are significantly reduced across females and males, and the differences between the sexes are likely to be small.

4.2 *Effect of Age, Sex, and Side of Surgery on Ischemia*

The results of the logistic regression analyses are presented in Table 3. The age distribution was as follows: 1% between 40 and 50 years old, 11% between 50 and 60 years old, 34% between 60 and 70 years old, 37% between 70 and 80 years old, 16% between 80 and 90 years old, and 1% between 90 and 100 years old. The odds of an ischemic-change increase by 1.03 times for each 10-year increase in age and is

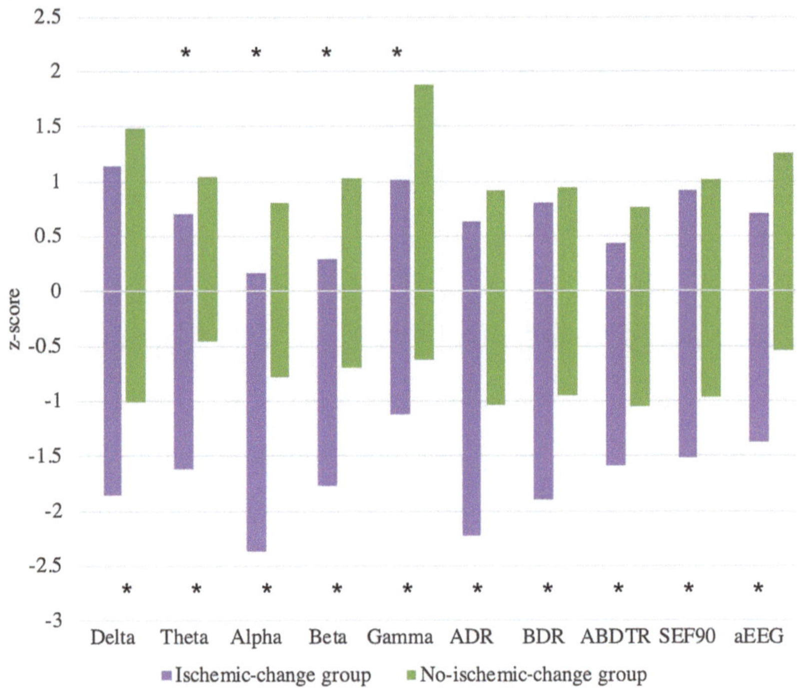

Fig. 10 Comparison of median *z*-scores for ipsilateral QEEG parameters between ischemic-change and no-ischemic-change groups in males. Positive *z*-scores denote peak values, whereas negative *z*-scores denote trough values

Table 3 Coefficients of logistic regression for age, sex, and side of surgery

Age	Sex	Side of surgery
1.03 (1.01–1.06) $p = 0.010$*	1.98 (1.26–3.11) $p = 0.003$*	1.17 (0.75–1.84) $p = 0.483$

The 95% confidence intervals are included in parentheses
*$p < 0.05$

statistically significant ($p < 0.05$), and the odds of an ischemic change for women are 1.98 times those for men, which is also statistically significant ($p < 0.05$). The odds of an ischemic change are 1.17 times greater for right-sided CEA than for left-sided CEA; however, this difference is not statistically significant at the 0.05 level [47]. In a systematic review, Leung et al. found a modest increase in perioperative risk with age in CEA [48], which supports our finding that advancing age is associated with a small increase in the risk of EEG ischemic change during CEA. Our results indicate that that females have a higher risk of ischemic change during CEA than males; this finding is supported by a retrospective analysis performed by Dansey et al. who found that female CEA patients have higher perioperative risk [49]. This disparity may be partially explained by Lee et al.'s systematic review, which found that female patients who had undergone vascular surgery had more severe disease [50]. Our results did not show a significant difference in ischemia between left-sided CEA and right-sided CEA. Uncertainty exists as to whether the side of the CEA influences the rate of complications, although Girard et al. found that left-side CEA is associated with increased postoperative complications [51].

5 Summary

In the operating room and the intensive care unit, cEEG is increasingly employed for real-time monitoring of brain activity, notably for identifying and monitoring cerebral ischemia. Because prolonged visual analysis of the cEEG is tedious, modern EEG software packages compute and display several QEEG parameters useful in the detection of cerebral ischemia.

Multiple studies have found a correlation between QEEG parameters and cerebral ischemia, but such studies have been conducted using a limited number of patients. We investigated the utility of ten QEEG parameters in detecting cerebral ischemia during CEA using a large database of patients who were monitored with intraoperative EEG. Our results show that decreases in the investigated QEEG parameters were associated with visual EEG changes indicative of post-clamp ischemia. When compared to the pre-clamp baseline, all ten ipsilateral QEEG parameters decreased significantly relative to the pre-clamp baseline, while contralateral QEEG parameters decreased less. In addition, our results indicate that a decrease in EEG amplitude is more prevalent in ischemia than an increase in activity at lower frequencies. These results are consistent with prior studies [34, 35]. Other findings include that the risk of EEG ischemic change increases with age, and females have a higher risk of EEG ischemic change during CEA than males.

Our work contains several limitations. One limitation is that the QEEG correlations were established for the average of each condition because the analyses were based on a patient's 10-minute cEEG recording that was labeled as ischemic or not. This may partly explain why all ten QEEG parameters showed similar behavior in ischemia. Further research is required to characterize the QEEG correlations at the 20-second interval and evaluate their performance in statistical models for predicting ischemia. Another limitation is that the patients in our database were drawn from a single healthcare system; patients from multiple hospitals would provide more diversity. Finally, the EEGs were evaluated for ischemia by a single expert reviewer.

The utility of some QEEG parameters for characterizing ischemia is promising; however, additional research is required to evaluate their applicability in statistical and machine learning algorithms that can accurately predict cerebral ischemia.

Acknowledgments This work was supported by the National Library of Medicine of the National Institutes of Health under award number T15 LM007059 and by the National Institute of General Medical Sciences of the National Institutes of Health under award number T32 GM008208. The content is solely the responsibility of the authors and does not necessarily represent the official views of the National Institutes of Health.

References

1. Smilowitz NR, Gupta N, Ramakrishna H, Guo Y, Berger JS, Bangalore S. Perioperative major adverse cardiovascular and cerebrovascular events associated with noncardiac surgery. JAMA Cardiol. 2017;2(2):181–7.
2. Al-Hader R, Al-Robaidi K, Jovin T, Jadhav A, Wechsler LR, Thirumala PD. The incidence of perioperative stroke: estimate using state and national databases and systematic review. J Stroke. 2019;21(3):290–301.
3. Mehta A, Gleason T, Wechsler L, Winger D, Wang L, Thirumala PD. Perioperative stroke as a predictor of mortality and morbidity in patients undergoing CABG. J Clin Neurosci. 2017;44:175–9.
4. Salazar JD, Wityk RJ, Grega MA, Borowicz LM, Doty JR, Petrofski JA, et al. Stroke after cardiac surgery: short- and long-term outcomes. Ann Thorac Surg. 2001;72(4):1195–201.
5. Ko SB. Perioperative stroke: pathophysiology and management. Korean J Anesthesiol. 2018;71(1):3–11.
6. Sacco RL, Kasner SE, Broderick JP, Caplan LR, Connors JJ, Culebras A, et al. An updated definition of stroke for the 21st century: a statement for healthcare professionals from the American Heart Association/American Stroke Association. Stroke. 2013;44(7):2064–89.
7. Sharbrough FW, Messick JM, Jr., Sundt TM, Jr. Correlation of continuous electroencephalograms with cerebral blood flow measurements during carotid endarterectomy. Stroke. 1973;4(4):674–83.
8. Hossmann KA. Viability thresholds and the penumbra of focal ischemia. Ann Neurol. 1994;36(4):557–65.
9. Jordan KG. Emergency EEG and continuous EEG monitoring in acute ischemic stroke. J Clin Neurophysiol. 2004;21(5):341–52.
10. Astrup J, Symon L, Branston NM, Lassen NA. Cortical evoked potential and extracellular K+ and H+ at critical levels of brain ischemia. Stroke. 1977;8(1):51–7.
11. Pinkerton JA, Jr. EEG as a criterion for shunt need in carotid endarterectomy. Ann Vasc Surg. 2002;16(6):756–61.
12. Chiappa KH, Burke SR, Young RR. Results of electroencephalographic monitoring during 367 carotid endarterectomies. Use of a dedicated minicomputer. Stroke. 1979;10(4):381–8.
13. Anetakis KM, Dolia JN, Desai SM, Balzer JR, Crammond DJ, Thirumala PD, et al. Last electrically well: intraoperative neurophysiological monitoring for identification and triage of large vessel occlusions. J Stroke Cerebrovasc Dis. 2020;29(10):105158.
14. Curtis JA, Johansen K. Techniques in carotid artery surgery. Neurosurg Focus. 2008;24(2):E18.
15. Thirumala PD, Thiagarajan K, Gedela S, Crammond DJ, Balzer JR. Diagnostic accuracy of EEG changes during carotid endarterectomy in predicting perioperative strokes. J Clin Neurosci. 2016;25:1–9.
16. Chang R, Reddy RP, Sudadi S, Balzer J, Crammond DJ, Anetakis K, et al. Diagnostic accuracy of various EEG changes during carotid endarterectomy to detect 30-day perioperative stroke: A systematic review. Clin Neurophysiol. 2020;131(7):1508–16.
17. Liu H, Di Giorgio AM, Williams ES, Evans W, Russell MJ. Protocol for electrophysiological monitoring of carotid endarterectomies. J Biomed Res. 2010;24(6):460–6.
18. Klem GH, Lüders HO, Jasper HH, Elger C. The ten-twenty electrode system of the International Federation. The International Federation of Clinical Neurophysiology. Electroencephalogr Clin Neurophysiol Suppl. 1999;52:3–6.
19. Sutter R, Kaplan PW. The history of continuous EEG monitoring. Continuous EEG Monitoring: Principles and Practice. 2017:3–12.
20. Herman ST, Abend NS, Bleck TP, Chapman KE, Drislane FW, Emerson RG, et al. Consensus statement on continuous EEG in critically ill adults and children, part I: indications. J Clin Neurophysiol. 2015;32(2):87–95.

21. Herman ST, Abend NS, Bleck TP, Chapman KE, Drislane FW, Emerson RG, et al. Consensus statement on continuous EEG in critically ill adults and children, part II: personnel, technical specifications, and clinical practice. J Clin Neurophysiol. 2015;32(2):96–108.
22. Sun Y, Wei C, Cui V, Xiu M, Wu A. Electroencephalography: Clinical Applications During the Perioperative Period. Front Med (Lausanne). 2020;7:251.
23. Ng MC, Jing J, Westover MB. A Primer on EEG Spectrograms. J Clin Neurophysiol. 2022;39(3):177–83.
24. Babadi B, Brown EN. A review of multitaper spectral analysis. IEEE Trans Biomed Eng. 2014;61(5):1555–64.
25. Purdon PL, Sampson A, Pavone KJ, Brown EN. Clinical electroencephalography for anesthesiologists: Part I: background and basic signatures. Anesthesiology. 2015;123(4):937–60.
26. Zanner R, Pilge S, Kochs EF, Kreuzer M, Schneider G. Time delay of electroencephalogram index calculation: analysis of cerebral state, bispectral, and Narcotrend indices using perioperatively recorded electroencephalographic signals. Br J Anaesth. 2009;103(3):394–9.
27. Rampil IJ. A primer for EEG signal processing in anesthesia. Anesthesiology. 1998;89(4):980–1002.
28. Husain AM, Sinha SR. Continuous EEG monitoring: Principles and Practice: Springer; 2017.
29. Hammond DC, Walker J, Hoffman D, Lubar JF, Trudeau D, Gurnee R, et al. Standards for the use of quantitative electroencephalography (QEEG) in neurofeedback: A position paper of the International Society for Neuronal Regulation. Journal of Neurotherapy. 2004;8(1):5–27.
30. Haneef Z, Levin HS, Frost JD, Jr., Mizrahi EM. Electroencephalography and quantitative electroencephalography in mild traumatic brain injury. J Neurotrauma. 2013;30(8):653–6.
31. Livint Popa L, Dragos H, Pantelemon C, Verisezan Rosu O, Strilciuc S. The role of quantitative EEG in the diagnosis of neuropsychiatric disorders. J Med Life. 2020;13(1):8–15.
32. Freye E, Levy JV. Cerebral monitoring in the operating room and the intensive care unit: an introductory for the clinician and a guide for the novice wanting to open a window to the brain. Part I: The electroencephalogram. J Clin Monit Comput. 2005;19(1–2):1–76.
33. Blume WT, Ferguson GG, McNeill DK. Significance of EEG changes at carotid endarterectomy. Stroke. 1986;17(5):891–7.
34. Kamitaki BK, Tu B, Wong S, Mendiratta A, Choi H. Quantitative EEG changes correlate with post-clamp ischemia during carotid endarterectomy. J Clin Neurophysiol. 2021;38(3):213–20.
35. Foreman B, Claassen J. Quantitative EEG for the detection of brain ischemia. Crit Care. 2012;16(2):216.
36. Rosenthal ES, Biswal S, Zafar SF, O'Connor KL, Bechek S, Shenoy AV, et al. Continuous electroencephalography predicts delayed cerebral ischemia after subarachnoid hemorrhage: a prospective study of diagnostic accuracy. Ann Neurol. 2018;83(5):958–69.
37. Tian J, Liu LD, Zhou Y, Zhang Z, Pu YH, Liu DC, et al. The change index of quantitative electroencephalography for evaluating the prognosis of large hemispheric infarction. J Integr Neurosci. 2021;20(2):341–7.
38. Jildenstål P, Bäckström A, Hedman K, Warrén-Stomberg M. Spectral edge frequency during general anaesthesia: a narrative literature review. J Int Med Res. 2022;50(8):3000605221118682.
39. Bruns N, Felderhoff-Müser U, Dohna-Schwake C. aEEG as a useful tool for neuromonitoring in critically ill children – current evidence and knowledge gaps. Acta Paediatr. 2021;110(4):1132–40.
40. Sheorajpanday RV, Nagels G, Weeren AJ, van Putten MJ, De Deyn PP. Reproducibility and clinical relevance of quantitative EEG parameters in cerebral ischemia: a basic approach. Clin Neurophysiol. 2009;120(5):845–55.
41. Claassen J, Hirsch LJ, Kreiter KT, Du EY, Connolly ES, Emerson RG, et al. Quantitative continuous EEG for detecting delayed cerebral ischemia in patients with poor-grade subarachnoid hemorrhage. Clin Neurophysiol. 2004;115(12):2699–710.
42. Halliday A, Harrison M, Hayter E, Kong X, Mansfield A, Marro J, et al. 10-year stroke prevention after successful carotid endarterectomy for asymptomatic stenosis (ACST-1): a multicentre randomised trial. Lancet. 2010;376(9746):1074–84.

43. Visser GH, Wieneke GH, van Huffelen AC. Carotid endarterectomy monitoring: patterns of spectral EEG changes due to carotid artery clamping. Clin Neurophysiol. 1999;110(2):286–94.
44. De Almeida KT, Simon MV, Chiappa KH. EEG monitoring during carotid endarterectomy. Intraoperative Neurophysiology: A Comprehensive Guide to Monitoring and Mapping 2018. p. 95.
45. Laman DM, Wieneke GH, van Duijn H, Veldhuizen RJ, van Huffelen AC. QEEG changes during carotid clamping in carotid endarterectomy: spectral edge frequency parameters and relative band power parameters. J Clin Neurophysiol. 2005;22(4):244–52.
46. Pedapati V, Du K, Mina A, Bradley A, Espino J, Batmanghelich K, et al. Quantitative EEG changes in carotid endarterectomy correlated with ischemia. 2022 IEEE Signal Processing in Medicine and Biology Symposium (SPMB). 2022:1–5.
47. Du K, Pedapati V, Mina A, Bradley A, Espino J, Batmanghelich K, et al. EEG changes correlated with ischemia across the sexes in carotid endarterectomy. 2022 IEEE Signal Processing in Medicine and Biology Symposium (SPMB). 2022:1–4.
48. Leung YYR, Bera K, Urriza Rodriguez D, Dardik A, Mas JL, Simonte G, et al. Safety of carotid endarterectomy for symptomatic stenosis by age: meta-analysis with individual patient data. Stroke. 2023;54(2):457–67.
49. Dansey KD, Pothof AB, Zettervall SL, Swerdlow NJ, Liang P, Schneider JR, et al. Clinical impact of sex on carotid revascularization. J Vasc Surg. 2020;71(5):1587–94.e2.
50. Lee MH, Li PY, Li B, Shakespeare A, Samarasinghe Y, Feridooni T, et al. A systematic review and meta-analysis of sex- and gender-based differences in presentation severity and outcomes in adults undergoing major vascular surgery. J Vasc Surg. 2022;76(2):581–94.e25.
51. Girard LP, Feasby TE, Eliasziw M, Quan H, Kennedy J, Barnett HJ, et al. Complication rates after left- versus right-sided carotid endarterectomy. Circ Cardiovasc Qual Outcomes. 2009;2(6):642–7.

Calibration Methods for Automatic Seizure Detection Algorithms

Ana Borovac, David Hringur Agustsson, Tomas Philip Runarsson, and Steinn Gudmundsson

1 Introduction

Seizures are a common neurological emergency, with an estimated prevalence of 1% [30], which can cause permanent brain damage, and even death, if untreated [45]. Almost half of the neonates affected by seizures face long-term neurodevelopmental disorders [51]. Adults experiencing seizures are at higher risk for psychiatric disorders such as depression and are about 10 times more likely to commit suicide than the general population [22]. Heart dysfunction provoked by seizures can cause sudden death [44]. To improve the lives of people with seizures, prompt detection and appropriate treatment are crucial. Treatment options include anti-epileptic drugs, brain surgery, and electrical brain stimulation [27, 28, 38, 41].

The current gold standard for neonatal and adult seizure detection is an electroencephalogram (EEG), a recording of the electrical activity of the brain. EEG signal acquisition is typically done by placing electrodes on the scalp, and the voltage difference between pairs of electrodes is recorded. The recordings can span from minutes to days, depending on the clinical indication for EEG monitoring. Due to signal complexity and high variability [20, 34, 53], analysis of EEG recordings requires time and special expertise that is not always available [7].

A. Borovac
Faculty of Industrial Engineering, Mechanical Engineering and Computer Science, University of Iceland, Reykjavik, Iceland

Kvikna Medical ehf., Reykjavik, Iceland
e-mail: anb48@hi.is

D. H. Agustsson · T. P. Runarsson · S. Gudmundsson (✉)
Faculty of Industrial Engineering, Mechanical Engineering and Computer Science, University of Iceland, Reykjavik, Iceland
e-mail: dha4@hi.is; tpr@hi.is; steinng@hi.is

A. Ahmed, J. Picone (eds.), *Machine Learning Applications in Medicine and Biology*,
https://doi.org/10.1007/978-3-031-51893-5_3

The goal of automated seizure detection algorithms (SDAs) [35, 43] is to accelerate EEG analysis significantly while preserving the current level of diagnostic accuracy. Such SDAs could enable the widespread use of EEGs, e.g. in intensive care units, without the need for experts to monitor each recording. The development of SDAs that perform as well as human experts faces two main challenges. First, due to patient privacy issues, there is often a limited amount of data available for algorithm training [10]. Second, obtaining precise annotations of seizure onset and offset times is challenging, as human experts may disagree on the presence of seizure events [9, 17]. SDAs trained on relatively small datasets are expected to have difficulties classifying unseen EEG segments accurately. Compounding the problem is the presence of label noise in the data because of ambiguity in the human annotations of the EEG [5]. Combining seizure/non-seizure predictions with confidence estimates would make the detectors more useful in a clinical setting [3, 24]. By doing so, EEG intervals with low-confidence predictions can be flagged for review by a human expert. The end result would be faster analysis without compromising the accuracy of the annotations. For example, a study using an SDA based on a support vector machine (SVM) suggests that by passing 40% of the data with the least confident predictions to a human expert, an accuracy of 99% could be achieved [2].

Many modern SDAs are based on deep neural networks (DNNs) [35, 43]. For a given EEG segment, a neural network classifier outputs a value between 0 and 1. This value can be interpreted as an estimate of the probability that the segment contains a seizure. A value close to 0 indicates that the segment is unlikely to contain a seizure, and a value close to 1 indicates that the segment most likely contains a seizure. The value can thus be regarded as the *confidence* the classifier has in the prediction. By thresholding at, say, 0.5, the segment can be classified as a seizure or non-seizure segment and labeled accordingly in the EEG recording. However, the accuracy of these confidence estimates has received limited attention in the context of SDAs [6]. In case the estimates are accurate, a classifier is considered to be well-*calibrated.* In other words, if a classifier outputs a confidence estimate of, e.g., 0.7 for some EEG segments, and it is correct in its prediction for 70% of these segments, the classifier is well-calibrated. The same should also hold for other confidence levels.

Guo et al. [16] have reported that DNNs trained on image and document classification tasks tend to be overconfident in their predictions, despite achieving high classification accuracy. Based on their empirical results, they suggest several potential causes that result in poorly calibrated DNNs, including increased model capacity, batch normalization, training with small weight decay, and using the cross-entropy loss function [54]. Thulasidasan et al. [50] suggest that training with 0/1 annotations negatively influences calibration and that it is improved by utilizing mixup [56]. Hein et al. [19] showed that DNNs employing the ReLU activation function can be particularly overconfident in their predictions for data far away from the training data. On the other hand, Minderer et al. [31] found that state-of-the-art DNNs for image classification tend to be well-calibrated and suggest that improvements in model accuracy benefit calibration. It should be noted that the image classification datasets employed in the above studies typically feature hundreds of classes, whereas seizure detection is normally formulated as a binary classification task. Researchers have proposed various approaches to improve the calibration of DNNs [1, 12]. These methods include post-processing techniques such as isotonic regression [55] and Platt scaling [39], which adjust the output probabilities of the network in order to improve calibration. Methods such as mixup [50, 56] and dropout [11] modify the training process, and in the case of dropout, also the prediction process.

In this work, we extend our previous analysis of SDA calibration [6] by analyzing four different calibration methods that have been found to work well with DNNs, albeit in different settings. We show that neonatal and adult SDAs based on a convolution neural network are overconfident in their predictions and that detectors with higher overall accuracy are better calibrated. All the calibration methods evaluated here, temperature scaling, ensemble, dropout, and mixup, make the detector less confident for incorrect predictions. A comparison of the methods is done on two publicly available datasets: one adult and one neonatal dataset.

2 Methods

2.1 Data

The adult EEG dataset was obtained from version 2.0.0 of the TUH EEG seizure corpus [18], which consists of recordings with diverse recording setups and seizure types. The most frequent seizure type is focal nonspecific seizures, but other types, such as generalized nonspecific seizures and complex partial seizures are also present. In this study, we utilized a subset of recordings recorded with averaged reference, i.e. average potential of all the electrodes was used as a reference. The acquisition of the signals was done with a version of a NicoletOne EEG system (Natus, USA), and the sampling frequency was between 250 and 1000 Hz. Human experts annotated the recordings using a bipolar temporal central parasagittal montage with 22 channels (Fig. 1), and the same montage was employed in this study. Specifically, channels Fp1-F7, F7-T3, T3-T5, T5-O1, Fp2-F8, F8-T4, T4-T6, T6-O2, T3-C3, C3-Cz, Cz-C4, C4-T4, Fp1-F3, F3-C3, C3-P3, P3-O1, Fp2-F4, F4-C4, C4-P4, P4-O2, A1-T3, and T4-A2 were derived from the recorded signals. The dataset contains predefined training, validation, and test sets.

The neonatal dataset used in this study consists of 79 recordings [47]. Acquisition of the EEG signals was done with NicoletOne EEG system (Natus, USA), using 19 electrodes with the reference electrode located at the midline and the sampling frequency was 256 Hz. To annotate the recordings, three human experts utilized a bipolar longitudinal (double banana) montage with 18 channels. Schematic representation of the montage is given in Fig. 1. For this study, the same set of channels was used, including Fp2-F4, F4-C4, C4-P4, P4-O2, Fp1-F3, F3-C3, C3-

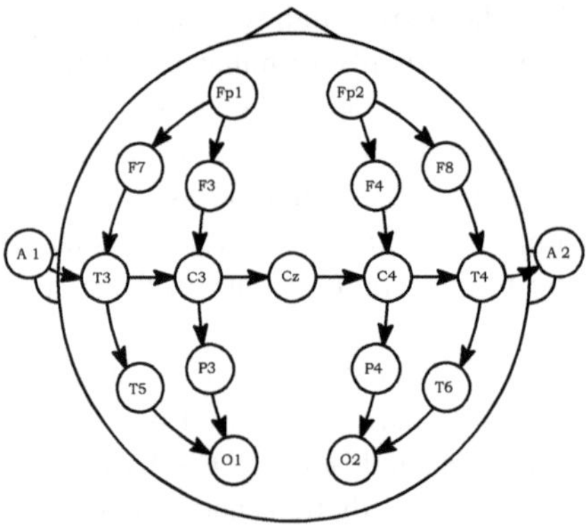

(a) Montage for adult EEG recordings.

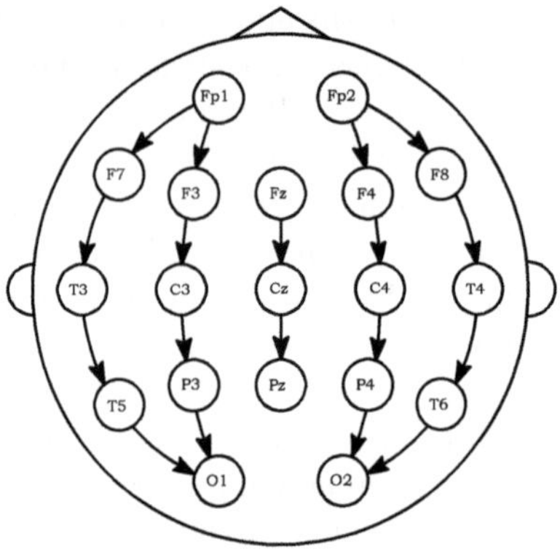

(b) Montage for neonatal EEG recordings.

Fig. 1 Montages for adult (**a**) and neonatal (**b**) EEG recordings. In both cases, the electrodes are positioned according to the 10–20 system. Each arrow denotes an EEG channel that is used as input to an SDA

Table 1 Summary statistics for the adult [18] and neonatal [47] datasets used in this study. A patient "with seizures" has at least one 16 seconds seizure segment. Standard deviations are shown in parentheses

	Adult data set			
	Training	Validation	Test	Neonatal data set
Number of patients	297	41	41	79
Total duration of recordings [hours]	603.08	372.21	119.98	111.90
Total duration of seizures [hours]	26.41	10.91	7.57	10.91
Fraction of seizure activity [%]	4.38	2.93	6.31	9.75
Average duration of recordings	2.03	9.08	2.93	1.42
per patient [hours]	(3.29)	(17.85)	(2.07)	(0.56)
Average duration of seizures	0.24	0.34	0.22	0.28
per patient with seizures [hours]	(0.41)	(0.42)	(0.31)	(0.38)
Number of seizure segments	19,148	8066	5197	8563
Number of non-seizure segments	127,220	79,759	24,547	20,233

P3, P3-O1, Fp2-F8, F8-T4, T4-T6, T6-O2, Fp1-F7, F7-T3, T3-T5, T5-O1, Fz-Cz, and Cz-Pz, which were derived from the recorded EEG signals. The dataset does not come with predefined training, validation, and test set splits. Table 1 shows summary statistics for the two datasets. We note that the datasets are imbalanced, i.e., less than 10% of the total recording duration corresponds to seizure segments.

Since most adult seizures are present in the frequency range between 3 and 30 Hz [15] and neonatal seizures can be as slow as 0.5 Hz [13], the EEG signals were filtered with a Butterworth band-pass filter with cutoff frequencies 0.5 and 30 Hz. Before filtering, the EEG signals were downsampled to 250 Hz and further downsampled to 62 Hz after filtering, reducing the input size and subsequently model size by approximately factor 4. After filtering and downsampling, each recording was cut into 16 seconds segments. To increase the amount of seizure data available for training, an overlap of 12 seconds was used for the seizure segments.

2.2 Seizure Detection Algorithm

The detector takes multichannel EEG as input and outputs seizure/non-seizure probability estimates. This is accomplished by extracting features from each EEG channel via 11 convolutional layers with 32 filters of size 3×1, followed by batch normalization layers and ReLU activation functions [36]. Average pooling is applied before the fourth, seventh, and tenth convolutional layers. An attention layer is used to combine feature vectors extracted from individual EEG channels into one feature vector [21]. The classification part of the network is a fully connected layer that maps feature vectors of dimension 58 to two outputs (seizure/non-seizure) utilizing a softmax activation function to obtain values in (0, 1) that can be interpreted as class probabilities. With only 29, 964 learnable parameters, the SDA is practically tiny, compared to state-of-the-art networks used in natural language processing and computer vision. A benefit of using such a small network is that it can be deployed on devices with limited computation resources. A detailed description of the SDA is given in [4].

The adult and neonatal detectors were trained by optimizing the negative log-likelihood loss function with the Adam optimizer and a mini-batch size of 256. To address the imbalance between the number of available seizure and non-seizure segments in the training sets, each mini-batch contained 128 seizure and 128 non-seizure segments. One epoch corresponds to a single pass through all available seizure segments and an equal number of randomly selected non-seizure segments. The SDAs were trained for 50 epochs where the initial learning rate of 0.001 was halved every 20 epochs. The number of epochs and the learning rate decay were chosen so that the area under the curve computed on the adult validation set was maximized. The hyper-parameter values used in this experiment are similar to those of previous experiments [5] conducted on the neonatal dataset. We observed that the performance of the SDA is insensitive to small changes in hyper-parameter values.

2.3 Calibration Methods

2.4 Temperature Scaling

Platt scaling [39] is a generic method to transform classifier outputs to a probability distribution over classes. It was originally proposed for use with SVM classifiers and has previously been used with an SVM-based neonatal seizure detector to smooth classifier outputs and to aggregate predictions over multiple channels [48]. The method fits a parameterized sigmoid or softmax function to (unscaled) classifier outputs. In [16] a simplified version with one learnable parameter called *temperature scaling* was used to improve the calibration of neural networks trained on image and document data. In case the non-seizure class is denoted with 0 and the seizure class with 1, the calibrated seizure/non-seizure probability estimates are obtained as follows:

$$\hat{p}_j^{(i)} = \frac{\exp\left(z_j^{(i)}/\tau\right)}{\exp\left(z_0^{(i)}/\tau\right) + \exp\left(z_1^{(i)}/\tau\right)}; \quad i = 1, 2, \ldots, N, \; j = 0, 1, \tag{1}$$

where $z_c^{(i)}$ $(c = 0, 1)$ are the unscaled outputs of instance i and $\hat{p}_j^{(i)}$ is the calibrated probability of instance i belonging to class j. The value of the τ parameter is chosen based on a held-out validation set after the network has been trained to avoid overfitting [39]. In case $\tau = 1$ the calibrated probabilities are equal to the softmax outputs and when τ is large the probabilities approach $1/2$. Applying temperature scaling to the unscaled outputs of the network does not change the predicted seizure/non-seizure labels, the only difference is in the probability (confidence) estimates.

2.5 Dropout

Dropout is a simple and widely used regularization technique for improving the generalization of DNNs [46]. The idea behind dropout is to randomly drop nodes from the network with a fixed probability, p. This forces the network to learn more robust features that are not dependent on any single node which subsequently reduces overfitting. Dropout is usually only used during training, i.e., the full network is used to obtain predictions. Dropout can also be employed in the prediction phase (Monte Carlo dropout). In this setting, the final seizure/non-seizure prediction is obtained by averaging T softmax outputs. It has been shown empirically that Monte Carlo improves the calibration of DNNs [29] and can be interpreted as approximate Bayesian inference [11]. Dropout with $p = 0.1$ was used for the convolutional and attention layers and $p = 0.5$ for the fully connected layer [11, 46]. The average of $T = 10$ softmax predictions was used for final probability estimates (averaging over a larger number of predictions gave similar results, data not shown).

2.6 Deep Ensembles

An ensemble of multiple DNNs, referred to as *deep ensemble* in the following, has been shown to give small improvements in classification performance compared to the best individual model in the ensemble [23]. An added benefit of using an ensemble of DNNs is improved calibration. Lakshminarayanan et al. [26] found that an ensemble with only five models trained with the same setup can lead to a noticeable improvement in calibration. Here we used an ensemble of 10 SDAs. Each individual SDA was trained with the same training parameters and the same data. The resulting SDAs were nevertheless not identical since network weights were randomly initialized for each network prior to training. Additionally, the order in which the data were presented to the network was different since the data were randomly shuffled for every epoch. Once the detectors were trained, the final prediction was obtained by averaging the softmax outputs.

2.7 *Mixup*

Mixup is a data-agnostic augmentation method that has been found to improve the generalization of many neural network architectures [56]. It has also been found to improve the model calibration of classifiers for both images and text [50]. Mixup creates augmented training examples by forming linear combinations of feature-target pairs. A new feature-target $(\tilde{x}, \tilde{y})$ is generated as follows:

$$\tilde{x} = \lambda x^{(i)} + (1 - \lambda)x^{(j)}, \tag{2}$$

$$\tilde{y} = \lambda y^{(i)} + (1 - \lambda)y^{(j)}, \tag{3}$$

where $x^{(i)}$ and $x^{(j)}$ are two randomly selected training EEG segments, $y^{(i)}$ and $y^{(j)}$ are corresponding 0/1 (non-seizure/seizure) labels and $\lambda \in [0, 1]$ is a random variable drawn from a Beta distribution with hyper-parameter α. It is important to select an appropriate α to achieve good results, in this study, $\alpha = 0.3$ was used after testing several different values on the adult validation set.

2.8 *Evaluation*

The adult SDA was assessed using a dedicated test set, while the evaluation of the neonatal SDA was done using leave-one-subject-out cross-validation since the dataset lacks a distinct test set.

The SDAs were assessed for their classification performance using the area under the curve (AUC), sensitivity (SE), and specificity (SP). Sensitivity refers to the fraction of correctly classified seizure segments, while specificity refers to the fraction of correctly classified non-seizure segments. The *confidence* of a prediction is the softmax output of the predicted seizure/non-seizure class, i.e., the class with the higher probability estimate. Since the threshold for seizure/non-seizure prediction is set at 0.5, the confidence estimates range between 0.5 and 1.0.

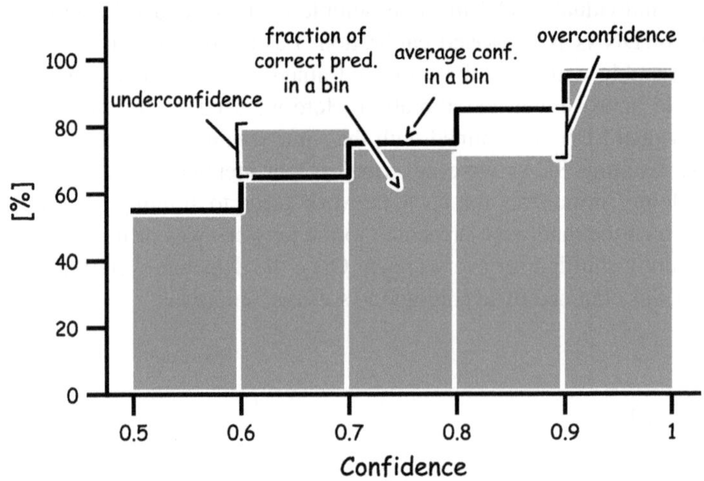

Fig. 2 Reliability diagram. The interval between the lowest (0.5) and highest (1.0) possible confidence values is split into five equally sized bins. All EEG segments are allocated to a bin based on the confidence of their predictions. Gray bars represent the fraction of the correctly predicted segments in a bin. The black curve represents the average confidence in each bin. Differences between the bars and the curve indicate miscalibration, i.e. the SDA is either underconfident or overconfident for the predictions in the bin

A *reliability diagram* [32] is a visual representation of a classifier's calibration, as shown in Fig. 2. The diagram shows the fraction of accurately predicted segments, providing an empirical estimation of the true underlying accuracy, against confidence levels. A well-calibrated classifier is indicated by empirical frequencies that align closely with the line of average confidence within a given bin. If there are sufficient data, the average confidence line should approximate the identity line.

To evaluate the calibration metrics, all N available seizure and non-seizure segments were split into $K = 5$ bins based on the confidence estimate made by the SDA. Bin edges were set such that the interval between the lowest (0.5) and highest (1.0) possible confidence is partitioned into equally sized intervals. The set of segments in bin k is denoted with B_k and $|B_k|$ is the number of segments in bin k. The fraction of correct predictions (*empirical frequency*) in a bin k is denoted with $\mathrm{acc}(B_k)$ and the average confidence estimate in bin k with $\mathrm{conf}(B_k)$.

The *expected calibration error* (ECE) [16],

$$\mathrm{ECE} = \sum_{k=1}^{K} \frac{|B_k|}{N} \, |\mathrm{acc}(B_k) - \mathrm{conf}(B_k)|, \tag{4}$$

measures the difference between predicted confidence and empirical frequency. Bins with more segments weigh more than bins with fewer segments. The closer the metric is to zero, the better calibrated the model is.

In medical applications, classifiers that are not overconfident in the predictions are preferred. Therefore, we include the *overconfidence error* (OE) [50] for calibration evaluation,

$$\mathrm{OE} = \sum_{k=1}^{K} \frac{|B_k|}{N} \mathrm{conf}(B_k) \cdot \max(\mathrm{conf}(B_k) - \mathrm{acc}(B_k), 0), \tag{5}$$

A modification of the static calibration error (SCE) [33] is proposed to capture calibration of individual classes (seizure and non-seizure) when the class frequencies differ widely,

$$\mathrm{SCE} = \frac{1}{C} \sum_{k=1}^{K} \sum_{c=1}^{C} \frac{|B_{c_k}|}{N_c} |\mathrm{acc}(B_{c_k}) - \mathrm{conf}(B_{c_k})|. \tag{6}$$

In comparison with the original definition in [33], this definition differs in the weighting factor $|B_{c_k}|/N_c$, where N_c is the number of segments of class c (seizure and non-seizure) and B_{c_k} is the set of segments of class c in bin k. Here the weights are proportional to the number of segments in each class and not the total number of segments. As a result, all the classes have the same weight in the overall sum and the imbalanced data issue is addressed. In other words, the static calibration error is the average expected calibration error using segments of just one class.

We also include two calibration metrics, which measure the distance of the estimated probability to the target label. Specifically, the Brier score (BS) [8],

$$\mathrm{BS} = \frac{1}{N} \sum_{i=1}^{N} \left(\hat{y}^{(i)} - y^{(i)}\right)^2, \tag{7}$$

and negative log-likelihood (NLL) [40],

$$\mathrm{NLL} = -\sum_{i=1}^{N} y^{(i)} \log \hat{y}^{(i)} + \left(1 - y^{(i)}\right) \log \left(1 - \hat{y}^{(i)}\right), \tag{8}$$

where $\hat{y}^{(i)}$ is the softmax output of instance i for class 1 (seizure class) and $y^{(i)}$ is the target label of instance i.

2.9 Implementation

All code was written in Python 3.9. EEG recordings in EDF format were read with the MNE library [14] (version 0.24.1) and pre-processed with SciPy [52] (version 1.8.0). The detectors were developed with PyTorch [37] (version 1.11.0) and an

NVIDIA GeForce GTX 1080 Ti graphics card. The code is available at a GitHub repository (github.com/anaborovac/Calibrated-SDA).

3 Results and Discussion

In the following, we refer to an SDA that does not utilize any specific calibration method as *uncalibrated*. We show that the calibration of detectors is highly correlated with overall classification accuracy as most correct and incorrect predictions appear to have high confidence. Using calibration methods does not result in perfectly calibrated SDAs; however, lower confidence in incorrect predictions is obtained with all the methods, temperature scaling, ensemble, dropout, and mixup.

3.1 Tuning Hyper-Parameters

The number of training epochs, learning decay schedule, and the α parameter in mixup were optimized by maximizing the AUC on the adult validation set. The neonatal dataset is relatively small, and it is therefore costly to set aside separate data for validation. Instead of reducing the amount of neonatal data available for training, we decided to simply train the neonatal detector using the same hyper-parameters that we obtained for the adult detector. Figure 3 shows the negative log-likelihood loss and AUC during training on the adult datasets. The validation loss fluctuates significantly, but the AUC is relatively stable after approx. 30 epochs. Longer training and different weight decay schedules gave similar results (data not shown). The fluctuations in the validation loss may be due to a small and imbalanced dataset, overfitting or ambiguity in annotations of seizure/non-seizure segments in the datasets.

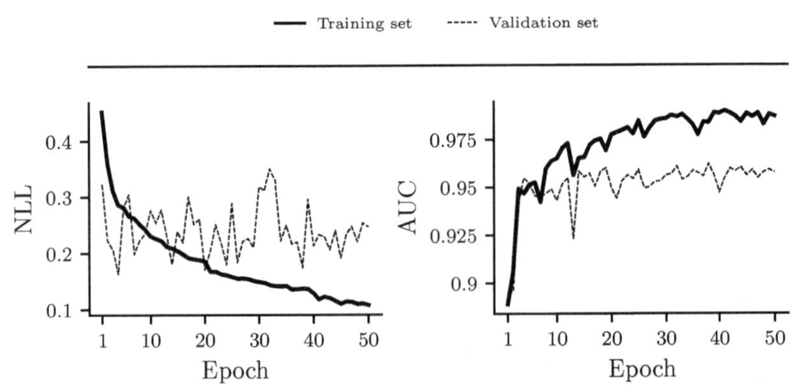

Fig. 3 Left: Negative log-likelihood (NLL) loss on the adult dataset during training of an uncalibrated SDA. Right: Corresponding area under the curve (AUC) values

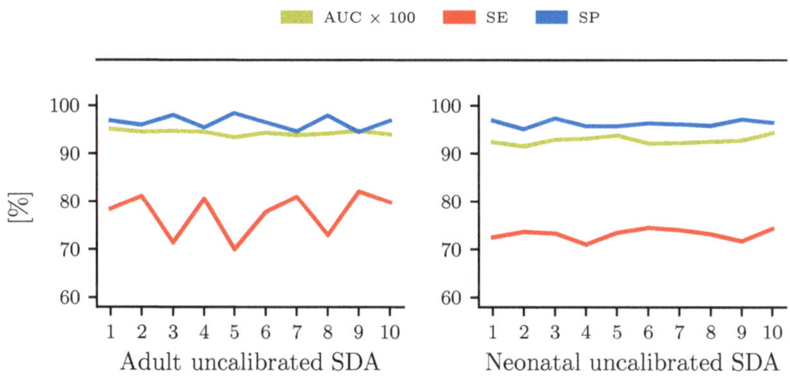

Fig. 4 Area under the curve (AUC), sensitivity (SE) and specificity (SP) for individual uncalibrated adult (left) and neonatal (right) SDAs from the deep ensembles (arbitrary order). Metrics are averaged across all patients, which have at least one seizure segment. A separate test set is used to compute metrics for the adult dataset while leave-one-subject-out cross-validation is used to compute the metrics for the neonatal dataset

3.2 Uncalibrated SDAs

To construct the deep ensembles, 10 sets of uncalibrated adult and neonatal SDAs were obtained by starting from random initial weights. We begin by analyzing these detectors individually to gain insight into the variability in the classification and calibration performance of individual classifiers.

Figure 4 shows that the performance of the adult SDAs is on average slightly better in comparison with the neonatal SDAs. This is not unexpected since the adult training set is significantly larger. The figure also shows that neonatal SDAs have less variability than adult SDAs. This may simply be a consequence of the use of leave-one-subject-out cross-validation on the neonatal dataset since averaging over 38 detectors has a smoothing effect. Figure 4 also shows the expected trade-off between sensitivity and specificity. Detectors with high seizure detection rates incorrectly classify more non-seizure segments as seizures and vice versa. However, since the AUC values are similar for all adult and neonatal detectors, respectively, the threshold for seizure/non-seizure prediction may be adjusted to achieve the desired classification performance. Adult SDAs exhibit considerable variance in sensitivity. A possible explanation is that the number of seizure segments available for training is much lower than the number of non-seizure segments. This may result in detectors that are not able to accurately capture the relevant features that differentiate seizures from non-seizure segments. Other factors that could contribute to higher variance are the heterogeneity of the different seizure types and incorrect annotations due to ambiguity in the scoring of EEGs by human experts [9, 17].

Figure 5 shows the expected and static calibration errors for individual adult and neonatal detectors. In both cases, the expected calibration and overconfidence errors were practically identical (data not shown). This means that the SDAs are

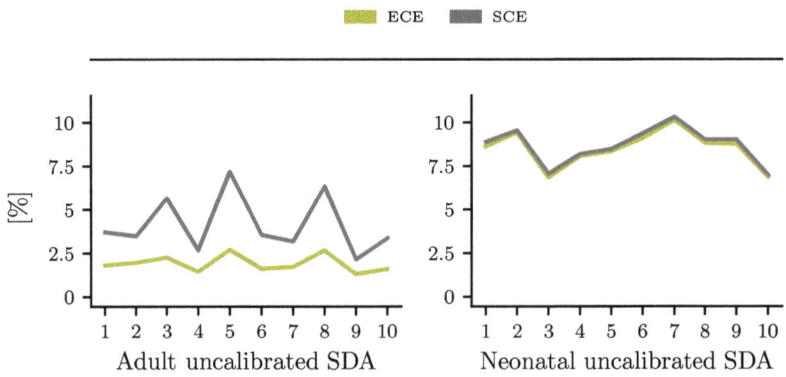

Fig. 5 Expected calibration error (ECE) and static calibration error (SCE) for individual uncalibrated adult (left) and neonatal (right) SDAs from the deep ensembles. The metrics are calculated based on all available segments in the test set for adult SDAs and left-out patients for neonatal SDAs

overconfident in their predictions, i.e. they are incorrect more frequently than the probabilities returned by the SDAs indicate. This can partly be explained by the use of ReLU activation functions [19] and batch normalization layers [16] in the detectors. The calibration may have been further compromised due to the training of the detectors using binary labels (non-seizure/seizure) [50] and cross-entropy loss function [54].

For the neonatal dataset, the expected and static calibration errors were very similar since the fraction of seizure segments in left-out neonatal patients is 49%, whereas for the adult test set, only 17% of the segments are seizure segments. Since static calibration error is an average of expected calibration errors calculated separately for seizure and non-seizure segments, it is more suitable for imbalanced datasets (e.g., the adult data set) than the expected calibration error.

Due to the imbalance in the datasets, we decided to investigate the calibration of seizure and non-seizure classes individually. Figure 6 shows that lower sensitivity/specificity results in higher expected calibration error, i.e. in worse calibration. This is a consequence of most (above 83%) segments being predicted with confidence greater than 0.9. From Eq. (4), it follows that the bin with segments predicted with the highest confidence affects the expected calibration error the most. Therefore, if the overall accuracy is very high and also close to the overall confidence, the detector is well-calibrated. The figure shows that the expected calibration error is higher for the seizure segments than for the non-seizure segments. However, there are more non-seizure segments in the adult test set, and therefore calibration on the non-seizure segments weighs more in the computation of the expected calibration error.

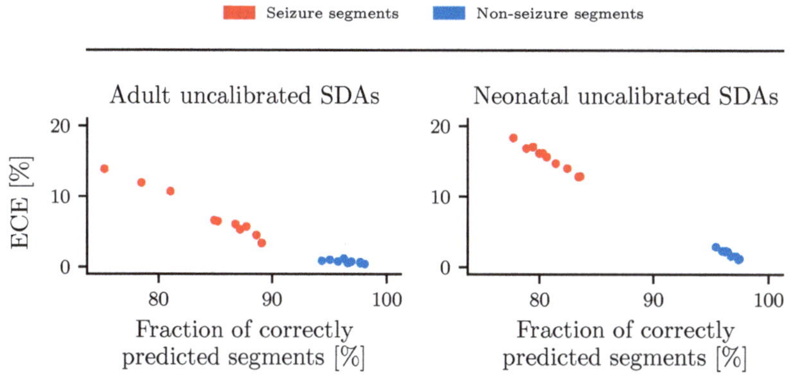

Fig. 6 Expected calibration error (ECE) calculated using seizure (red) and non-seizure (blue) segments. Each red and blue point represents one uncalibrated adult (left) and neonatal (right) SDA in the deep ensembles. The metrics are calculated from all available segments in the test set for the adult SDAs and from the left-out patients for neonatal SDAs

3.3 Calibrated SDAs

The AUC was averaged over all SDAs in an ensemble, and the individual classifier with AUC closest to the ensemble average was selected as a representative uncalibrated SDA in the following. Table 2 shows how different calibration methods affect the performance of the adult and neonatal SDAs. The classification performance of

Table 2 Patient-based classification metrics for uncalibrated and calibrated adult and neonatal SDAs. Metrics are averaged across patients with at least one 16 seconds long seizure segment. Standard deviations are shown in parentheses

		Calibrated		
	Uncalibrated	Ensemble	Dropout	Mixup
Adult SDA				
Area under the curve	0.94	0.95	0.95	0.96
	(0.11)	(0.11)	(0.11)	(0.09)
Sensitivity [%]	77.74	79.72	79.57	77.77
	(26.51)	(25.94)	(26.26)	(23.62)
Specificity [%]	96.44	97.51	97.0	96.45
	(5.44)	(4.07)	(4.14)	(5.03)
Neonatal SDA				
Area under the curve	0.93	0.95	0.92	0.92
	(0.12)	(0.10)	(0.14)	(0.13)
Sensitivity [%]	71.71	74.11	71.67	68.94
	(30.77)	(27.84)	(28.68)	(31.48)
Specificity [%]	97.11	97.28	94.07	96.91
	(7.04)	(7.95)	(12.77)	(5.98)

the adult SDAs using temperature scaling is identical to the uncalibrated classifier since the scaling procedure only affects confidence estimates and not the predicted class. The method is therefore not listed separately in the table. Temperature scaling was not applied to the neonatal SDAs since there was no dedicated validation set available for tuning the temperature parameter τ.

The performance metrics in Table 2 have fairly large variance across patients for all the SDAs, the sensitivity metric in particular. A likely explanation is that there are far fewer seizure segments (13%) in the training set, in comparison to non-seizure segments (Table 1). Although each mini-batch is balanced during training, the seizure class has fewer examples defining it. Furthermore, the metrics are only based on 31 adult and 38 neonatal patients, respectively, and differences in performance for 2–3 patients can end up having a significant effect on the overall mean and standard deviation. In both datasets, there are approx. three patients that the SDAs consistently had problems classifying and result in sensitivity values below 50% and/or specificity below 90%. The heterogeneity of the different seizure types and ambiguity in EEG signals are likely to contribute to the variability as well. From Table 2, we conclude that calibration methods do not outperform uncalibrated detectors, but they also do not noticeably degrade classification performance in terms of the area under the curve, sensitivity, and specificity. This is in line with previous studies, which applied ensembling, dropout, and mixup for image classification [23, 25, 49, 50, 57].

The calibration metrics in Table 3 were computed by averaging over all available test segments, instead of first computing the corresponding metrics over the patients and then averaging. The reason is that the number of segments behind each patient varies widely. For some patients, there are fewer than 100 segments, and this causes difficulties when computing metrics based on confidence bins. Overall, large

Table 3 Segment-based calibration metrics for uncalibrated and calibrated adult and neonatal SDAs. The metrics were calculated on all available segments in a test set for the adult SDAs and on the left-out patients for the neonatal SDAs

		Calibrated			
	Uncalibrated	Temp. scaling	Ensemble	Dropout	Mixup
Adult SDA					
Expected cal. error [%]	1.61	2.58	0.62	1.58	4.05
Overconfidence error [%]	1.55	0.0	0.03	0.0	0.0
Static calibration error [%]	3.54	3.21	2.02	2.31	3.65
Brier score	0.04	0.04	0.03	0.04	0.04
Negative log-likelihood	0.17	0.17	0.13	0.14	0.16
Neonatal SDA					
Expected cal. error [%]	8.76	–	4.72	4.80	1.76
Overconfidence error [%]	8.76	–	4.72	4.80	1.51
Static calibration error [%]	9.01	–	5.40	5.27	7.62
Brier score	0.10	–	0.08	0.09	0.11
Negative log-likelihood	0.59	–	0.29	0.32	0.35

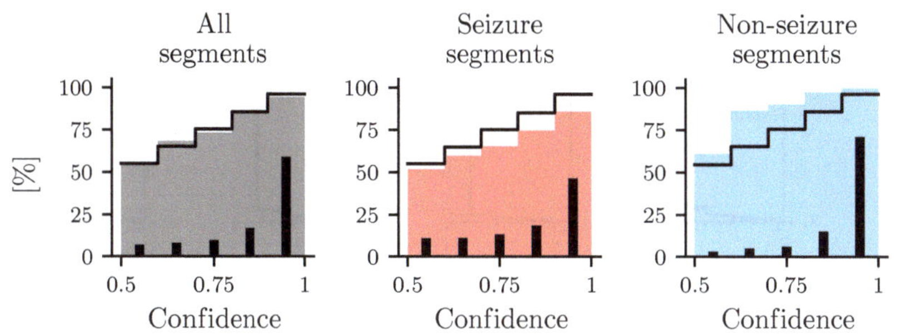

Fig. 7 Reliability diagrams for a neonatal SDA trained with mixup. The black step function indicates the average confidence of segments in each bin. Colored bars indicate the fraction of correctly predicted segments in a bin. Black bars indicate the fraction of segments in a bin. The detector is overconfident for seizure segments and underconfident for non-seizure segments leading to an expected calibration error of 1.76 and a static calibration error of 7.62

improvements in calibration were not observed for either dataset. However, we observe that all the calibration methods reduce the overconfidence error, a highly desired feature in medical applications. An overconfidence error close to zero for the adult data set implies that the calibrated SDAs are mainly underconfident in their predictions since the expected calibration errors are non-zero. The neonatal SDAs are overconfident, after employing the calibration methods, but the level of overconfidence error has decreased.

For the neonatal SDAs, a large difference between expected and static calibration errors was not expected since the dataset is balanced. For mixup, however, the two metrics differed considerably (Table 3). This indicates that the detector is overconfident for segments of one class and underconfident for the other. When the seizure and non-seizure components of the metrics are analyzed separately (Fig. 7), it appears that the SDA with mixup is overconfident in predicting seizure segments and not confident enough when predicting non-seizure segments.

The predicted confidence values are analyzed in more detail in Fig. 8 where the confidence estimates of correct and incorrect predictions are analyzed separately. Diagrams with similar patterns are obtained when only seizure or non-seizure segments are used (data not shown). The uncalibrated SDAs are confident in the predictions, both correct and incorrect, and most of them have confidence close to 1. As much as it is desired that the SDAs are confident in their correct predictions, it is also important that incorrect predictions are made with lower confidence, making it possible to inform the user that some parts of EEG are difficult to classify. In this case, binary seizure/non-seizure predictions can be accompanied by confidence values as illustrated in Fig. 9. For all the calibration methods, the number of incorrect predictions in the most confident bins is clearly reduced, which is what we want, but the SDAs are less confident in their correct predictions compared to uncalibrated SDAs. However, in the latter case, the reduction is fairly small and mainly the bin with the second largest confidence values increases in size.

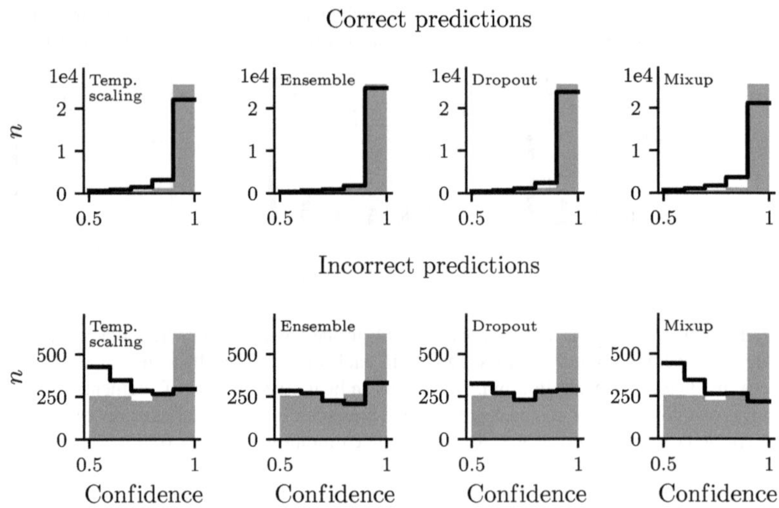

(a)

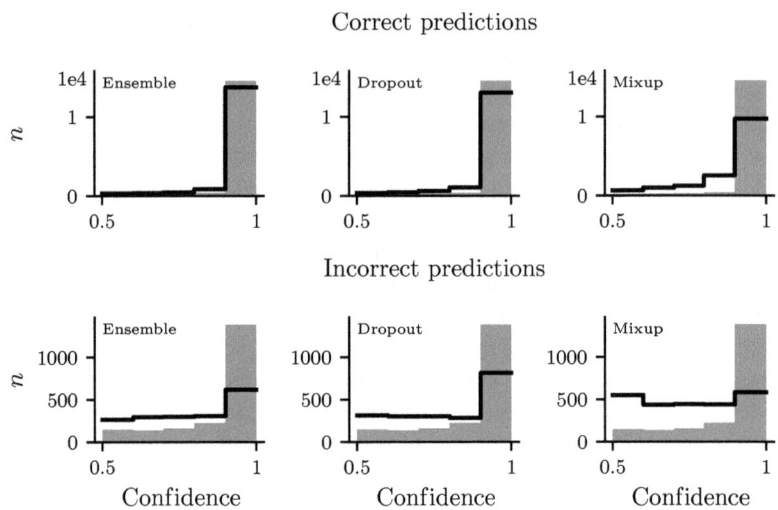

(b)

Fig. 8 Gray histograms represent the number of correctly and incorrectly classified segments of an arbitrarily chosen uncalibrated adult and neonatal SDA. The step functions represent the number of correctly and incorrectly classified segments. (**a**) Adult SDA. (**b**) Neonatal SDA

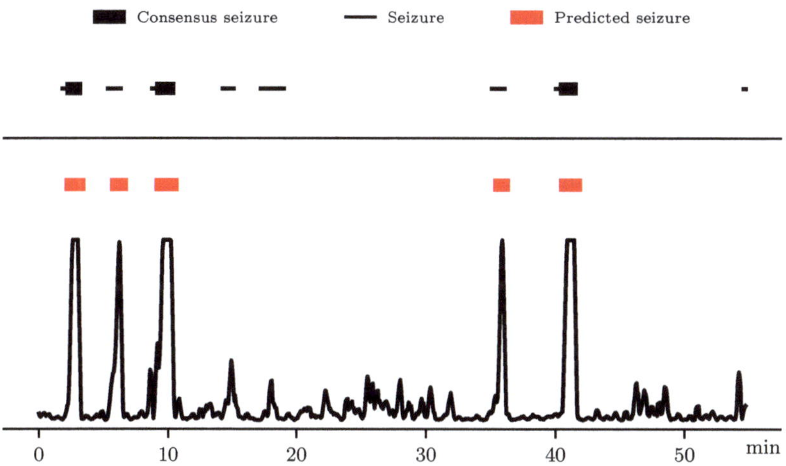

Fig. 9 Predictions for a short neonatal EEG recording (<1 hour), obtained with an SDA employing dropout. Black blocks represent consensus seizures where the three human experts scoring the recording were in agreement. Black lines represent seizures annotated by at least one of the three experts. Red blocks represent seizure predictions, and the corresponding probability estimates (confidence values) are denoted with a black curve

Mixup results in an SDA with the lowest average confidence among the calibration methods studied here and with the lowest number of segments in the most confident bin. This also means that the largest number of incorrect predictions are, as preferred, predicted with confidence close to 0.5. In the clinical setting, this would imply that more segments would be passed for an additional review done by a human expert, but only a few incorrectly classified would be missed. This is especially noticeable for neonatal SDA. Note that the hyper-parameters were tuned on the adult validation set and different results could be obtained if they were to be tuned on neonatal data.

4 Conclusion

In line with a previous study [6], we find that uncalibrated SDAs tend to be overconfident in their predictions, and the probability corresponding to an incorrect prediction gives little indication that the prediction is wrong. Since most predictions are made with confidence close to 1, a more accurate detector is also better calibrated.

The four calibration methods included in this study did not degrade classification performance, i.e. their sensitivity, specificity and AUC values were similar to the uncalibrated SDAs. A slight improvement among the classification metrics for the adult SDA utilizing ensembling, dropout, or mixup, was observed. These methods can be regarded as regularization techniques that reduce model overfitting and improve generalization, which in turn can explain increased detector accuracy.

With some additional computational costs, we found a modest improvement in calibration, with the ensemble method giving the largest improvement, followed by dropout. Mixup gave mixed results for the adult SDA but did better on the neonatal data. Temperature scaling gave little improvement.

In [6], we found that dropout gave a noticeable improvement in expected calibration error for both adult and neonatal, SDAs. A possible explanation for this discrepancy is that here we are starting from a more accurate uncalibrated classifier than in our earlier work, which then tends to be better calibrated [31]. The pre-processing and evaluation steps used here are slightly different from previous studies, which also contributes to the difference. In this study, the EEG data were not normalized prior to feeding it to the network, and the mini-batch size was larger. This resulted in slightly more accurate uncalibrated SDAs than before. Additionally, here the non-seizure segments do not overlap, and consequently, seizure segments represent a bigger portion of the test data. Since calibration on these segments tends to be worse than on the non-seizure segments, the overall expected calibration error is higher. Analyzing the calibration of each class is therefore advised in case of class imbalance.

All the calibrated detectors were noticeably less confident in incorrect predictions compared to uncalibrated detectors. EEG segments with low confidence values can then be passed to a human expert for manual review and eventual correction. This is a desirable property of an SDA if the main objective is to develop a detector that is as accurate as possible. However, in order for the SDA to be useful in clinical practice and make reviewing more time-efficient, the expert should not be given the majority of the acquired data for review. To limit the amount of data that requires human expertise, the confidence in correct predictions should be close to 1, and these EEG segments would therefore not be passed on to the expert. This pattern was observed for the ensemble and dropout, for temperature scaling and mixup, the drop in confidence for correct predictions was larger and unfavorable.

Further work is needed to evaluate if such a setup makes EEG scoring more efficient in the clinical environment. Introducing confidence estimates alongside binary seizure/non-seizure predictions to the EEG monitors could however confound the interpretation of the user. A study on how to best present the confidence estimates is therefore needed. Another possibility would be to present to the user only a few selected EEG segments from which the detector would learn and subsequently improve. The segments could, e.g., be chosen with an active learning approach [42].

References

1. Abdar, M., Pourpanah, F., Hussain, S., Rezazadegan, D., Liu, L., Ghavamzadeh, M., Fieguth, P., Cao, X., Khosravi, A., Acharya, U.R., et al.: A review of uncertainty quantification in deep learning: Techniques, applications and challenges. Information Fusion **76**, 243–297 (2021)

2. Becker, T., Vandecasteele, K., Chatzichristos, C., Van Paesschen, W., Valkenborg, D., Van Huffel, S., De Vos, M.: Classification with a deferral option and low-trust filtering for automated seizure detection. Sensors **21**(4), 1046 (2021)
3. Begoli, E., Bhattacharya, T., Kusnezov, D.: The need for uncertainty quantification in machine-assisted medical decision making. Nature Machine Intelligence **1**(1), 20–23 (2019)
4. Borovac, A., Gudmundsson, S., Thorvardsson, G., Moghadam, S.M., Nevalainen, P., Stevenson, N., Vanhatalo, S., Runarsson, T.P.: Ensemble learning using individual neonatal data for seizure detection. IEEE journal of translational engineering in health and medicine **10**, 1–11 (2022)
5. Borovac, A., Guðmundsson, S., Thorvardsson, G., Runarsson, T.P.: Influence of human-expert labels on a neonatal seizure detector based on a convolutional neural network. In: The NeurIPS 2021 Data-Centric AI Workshop (2021)
6. Borovac, A., Runarsson, T.P., Thorvardsson, G., Gudmundsson, S.: Calibration of Automatic Seizure Detection Algorithms. In: 2022 IEEE Signal Processing in Medicine and Biology Symposium (SPMB). pp. 1–6. IEEE (2022)
7. Boylan, G., Burgoyne, L., Moore, C., O'Flaherty, B., Rennie, J.: An international survey of EEG use in the neonatal intensive care unit. Acta paediatrica **99**(8), 1150–1155 (2010)
8. Brier, G.W., et al.: Verification of forecasts expressed in terms of probability. Monthly weather review **78**(1), 1–3 (1950)
9. Dereymaeker, A., Ansari, A.H., Jansen, K., Cherian, P.J., Vervisch, J., Govaert, P., De Wispelaere, L., Dielman, C., Matic, V., Dorado, A.C., et al.: Interrater agreement in visual scoring of neonatal seizures based on majority voting on a web-based system: the neoguard eeg database. Clinical Neurophysiology **128**(9), 1737–1745 (2017)
10. Eicher, J., Bild, R., Spengler, H., Kuhn, K.A., Prasser, F.: A comprehensive tool for creating and evaluating privacy-preserving biomedical prediction models. BMC Medical Informatics and Decision Making **20**(1), 1–14 (2020)
11. Gal, Y., Ghahramani, Z.: Dropout as a bayesian approximation: Representing model uncertainty in deep learning. In: international conference on machine learning. pp. 1050–1059. PMLR (2016)
12. Gawlikowski, J., Tassi, C.R.N., Ali, M., Lee, J., Humt, M., Feng, J., Kruspe, A., Triebel, R., Jung, P., Roscher, R., et al.: A survey of uncertainty in deep neural networks. arXiv preprint arXiv:2107.03342 (2021)
13. Gotman, J.: Automatic detection of epileptic seizures. Handbook of clinical neurophysiology **3**, 155–165 (2003)
14. Gramfort, A., Luessi, M., Larson, E., Engemann, D.A., Strohmeier, D., Brodbeck, C., Goj, R., Jas, M., Brooks, T., Parkkonen, L., et al.: MEG and EEG data analysis with MNE-Python. Frontiers in neuroscience p. 267 (2013)
15. Grewal, S., Gotman, J.: An automatic warning system for epileptic seizures recorded on intracerebral EEGs. Clinical neurophysiology **116**(10), 2460–2472 (2005)
16. Guo, C., Pleiss, G., Sun, Y., Weinberger, K.Q.: On calibration of modern neural networks. In: International conference on machine learning. pp. 1321–1330. PMLR (2017)
17. Halford, J., Shiau, D., Desrochers, J., Kolls, B., Dean, B., Waters, C., Azar, N., Haas, K., Kutluay, E., Martz, G., et al.: Inter-rater agreement on identification of electrographic seizures and periodic discharges in ICU EEG recordings. Clinical Neurophysiology **126**(9), 1661–1669 (2015)
18. Harati, A., Lopez, S., Obeid, I., Picone, J., Jacobson, M., Tobochnik, S.: The TUH EEG CORPUS: A big data resource for automated EEG interpretation. In: 2014 IEEE signal processing in medicine and biology symposium (SPMB). pp. 1–5. IEEE (2014)
19. Hein, M., Andriushchenko, M., Bitterwolf, J.: Why relu networks yield high-confidence predictions far away from the training data and how to mitigate the problem. In: Proceedings of the IEEE/CVF Conference on Computer Vision and Pattern Recognition. pp. 41–50 (2019)
20. Hrachovy, R.A., Mizrahi, E.M.: Atlas of neonatal electroencephalography. Springer Publishing Company (2015)

21. Isaev, D.Y., Tchapyjnikov, D., Cotten, C.M., Tanaka, D., Martinez, N., Bertran, M., Sapiro, G., Carlson, D.: Attention-based network for weak labels in neonatal seizure detection. Proceedings of machine learning research **126**, 479 (2020)
22. Jones, J.E., Hermann, B.P., Barry, J.J., Gilliam, F.G., Kanner, A.M., Meador, K.J.: Rates and risk factors for suicide, suicidal ideation, and suicide attempts in chronic epilepsy. Epilepsy & Behavior **4**, 31–38 (2003)
23. Ju, C., Bibaut, A., van der Laan, M.: The relative performance of ensemble methods with deep convolutional neural networks for image classification. Journal of Applied Statistics **45**(15), 2800–2818 (2018)
24. Kompa, B., Snoek, J., Beam, A.L.: Second opinion needed: communicating uncertainty in medical machine learning. NPJ Digital Medicine **4**(1), 1–6 (2021)
25. Krishnan, R., Tickoo, O.: Improving model calibration with accuracy versus uncertainty optimization. Advances in Neural Information Processing Systems **33**, 18237–18248 (2020)
26. Lakshminarayanan, B., Pritzel, A., Blundell, C.: Simple and scalable predictive uncertainty estimation using deep ensembles. Advances in neural information processing systems **30** (2017)
27. Lamberink, H.J., Otte, W.M., Bluemcke, I., Braun, K.P., Aichholzer, M., Amorim, I., Aparicio, J., Aronica, E., Arzimanoglou, A., Barba, C., et al.: Seizure outcome and use of antiepileptic drugs after epilepsy surgery according to histopathological diagnosis: a retrospective multicentre cohort study. The Lancet Neurology **19**(9), 748–757 (2020)
28. Le, V.T., Abdi, H.H., Sánchez, P.J., Yossef, L., Reagan, P.B., Slaughter, L.A., Firestine, A., Slaughter, J.L.: Neonatal antiepileptic medication treatment patterns: a decade of change. American journal of perinatology **38**(05), 469–476 (2021)
29. Leibig, C., Allken, V., Ayhan, M.S., Berens, P., Wahl, S.: Leveraging uncertainty information from deep neural networks for disease detection. Scientific reports **7**(1), 1–14 (2017)
30. Litt, B., Echauz, J.: Prediction of epileptic seizures. The Lancet Neurology **1**(1), 22–30 (2002)
31. Minderer, M., Djolonga, J., Romijnders, R., Hubis, F., Zhai, X., Houlsby, N., Tran, D., Lucic, M.: Revisiting the calibration of modern neural networks. Advances in Neural Information Processing Systems **34**, 15682–15694 (2021)
32. Niculescu-Mizil, A., Caruana, R.: Predicting good probabilities with supervised learning. In: Proceedings of the 22nd international conference on Machine learning. pp. 625–632 (2005)
33. Nixon, J., Dusenberry, M.W., Zhang, L., Jerfel, G., Tran, D.: Measuring Calibration in Deep Learning. In: CVPR Workshops. vol. 2 (2019)
34. Noachtar, S., Rémi, J.: The role of EEG in epilepsy: a critical review. Epilepsy & Behavior **15**(1), 22–33 (2009)
35. Olmi, B., Frassineti, L., Lanata, A., Manfredi, C.: Automatic Detection of Epileptic Seizures in Neonatal Intensive Care Units Through EEG, ECG and Video Recordings: A Survey. IEEE Access **9**, 138174–138191 (2021)
36. O'Shea, A., Lightbody, G., Boylan, G., Temko, A.: Investigating the impact of CNN depth on neonatal seizure detection performance. In: 2018 40th Annual International Conference of the IEEE Engineering in Medicine and Biology Society (EMBC). pp. 5862–5865. IEEE (2018)
37. Paszke, A., Gross, S., Massa, F., Lerer, A., Bradbury, J., Chanan, G., Killeen, T., Lin, Z., Gimelshein, N., Antiga, L., et al.: Pytorch: An imperative style, high-performance deep learning library. Advances in neural information processing systems **32** (2019)
38. Perucca, E., Brodie, M.J., Kwan, P., Tomson, T.: 30 years of second-generation antiseizure medications: impact and future perspectives. The Lancet Neurology **19**(6), 544–556 (2020)
39. Platt, J.: Probabilistic outputs for support vector machines and comparisons to regularized likelihood methods. Advances in large margin classifiers **10**(3), 61–74 (1999)
40. Quinonero-Candela, J., Rasmussen, C.E., Sinz, F., Bousquet, O., Schölkopf, B.: Evaluating predictive uncertainty challenge. In: Machine Learning Challenges Workshop. pp. 1–27. Springer (2006)
41. Razavi, B., Rao, V.R., Lin, C., Bujarski, K.A., Patra, S.E., Burdette, D.E., Geller, E.B., Brown, M.G.M., Johnson, E.A., Drees, C., et al.: Real-world experience with direct brain-responsive neurostimulation for focal onset seizures. Epilepsia **61**(8), 1749–1757 (2020)

42. Ren, P., Xiao, Y., Chang, X., Huang, P.Y., Li, Z., Gupta, B.B., Chen, X., Wang, X.: A survey of deep active learning. ACM computing surveys (CSUR) **54**(9), 1–40 (2021)
43. Saminu, S., Xu, G., Shuai, Z., Abd El Kader, I., Jabire, A.H., Ahmed, Y.K., Karaye, I.A., Ahmad, I.S.: A recent investigation on detection and classification of epileptic seizure techniques using EEG signal. Brain Sciences **11**(5), 668 (2021)
44. Schuele, S.U.: Effects of seizures on cardiac function. Journal of clinical neurophysiology **26**(5), 302–308 (2009)
45. Scott, R.C.: What are the effects of prolonged seizures in the brain? Epileptic Disorders **16**(s1), S6–S11 (2014)
46. Srivastava, N., Hinton, G., Krizhevsky, A., Sutskever, I., Salakhutdinov, R.: Dropout: a simple way to prevent neural networks from overfitting. The journal of machine learning research **15**(1), 1929–1958 (2014)
47. Stevenson, N.J., Tapani, K., Lauronen, L., Vanhatalo, S.: A dataset of neonatal EEG recordings with seizure annotations. Scientific data **6**, 190039 (2019)
48. Temko, A., Thomas, E., Marnane, W., Lightbody, G., Boylan, G.: EEG-based neonatal seizure detection with support vector machines. Clinical Neurophysiology **122**(3), 464–473 (2011)
49. Thagaard, J., Hauberg, S., Vegt, B.v.d., Ebstrup, T., Hansen, J.D., Dahl, A.B.: Can you trust predictive uncertainty under real dataset shifts in digital pathology? In: International Conference on Medical Image Computing and Computer-Assisted Intervention. pp. 824–833. Springer (2020)
50. Thulasidasan, S., Chennupati, G., Bilmes, J.A., Bhattacharya, T., Michalak, S.: On mixup training: Improved calibration and predictive uncertainty for deep neural networks. Advances in Neural Information Processing Systems **32** (2019)
51. Uria-Avellanal, C., Marlow, N., Rennie, J.M.: Outcome following neonatal seizures. In: Seminars in Fetal and Neonatal Medicine. vol. 18, pp. 224–232. Elsevier (2013)
52. Virtanen, P., Gommers, R., Oliphant, T.E., Haberland, M., Reddy, T., Cournapeau, D., Burovski, E., Peterson, P., Weckesser, W., Bright, J., et al.: SciPy 1.0: fundamental algorithms for scientific computing in Python. Nature methods **17**(3), 261–272 (2020)
53. Webb, L., Kauppila, M., Roberts, J.A., Vanhatalo, S., Stevenson, N.J.: Automated detection of artefacts in neonatal EEG with residual neural networks. Computer Methods and Programs in Biomedicine **208**, 106194 (2021)
54. Wei, H., Xie, R., Cheng, H., Feng, L., An, B., Li, Y.: Mitigating neural network overconfidence with logit normalization. In: International Conference on Machine Learning. pp. 23631–23644. PMLR (2022)
55. Zadrozny, B., Elkan, C.: Transforming classifier scores into accurate multiclass probability estimates. In: Proceedings of the eighth ACM SIGKDD international conference on Knowledge discovery and data mining. pp. 694–699 (2002)
56. Zhang, H., Cisse, M., Dauphin, Y.N., Lopez-Paz, D.: mixup: Beyond empirical risk minimization. arXiv preprint arXiv:1710.09412 (2017)
57. Zhang, Z., Dalca, A.V., Sabuncu, M.R.: Confidence calibration for convolutional neural networks using structured dropout. arXiv preprint arXiv:1906.09551 (2019)

48. Ren, P., Xiao, Y., Chang, X., Huang, P.Y., Li, Z., Gupta, B.B., Chen, X., Wang, X.: A survey of deep active learning. ACM computing surveys (CSUR) 54(9), 1–40 (2021)

[illegible]

Real-Time Wavelet Processing and Classifier Algorithms Enabling Single-Channel Diagnosis of Lower Urinary Tract Dysfunction

S. J. A. Majerus, M. Abdelhady, V. Abbaraju, J. Han, L. Brody, and M. Damaser

1 Urodynamic Testing for Urology Diagnosis

Lower urinary tract dysfunction (LUTD) affects nearly 20% of the US adult population and can severely reduce quality of life [1, 2]. Symptoms of LUTD include increased urinary urgency and urinary incontinence, which is the involuntary leakage of urine during the storage phase of the bladder [1, 3, 4]. Individuals

S. J. A. Majerus (✉)
Department of Electrical, Computer, and Systems Engineering, Case Western Reserve University, Cleveland, OH, USA

Advanced Platform Technology Center, Veterans Affairs Medical Center, Cleveland, OH, USA
e-mail: steve.majerus@case.edu

M. Abdelhady
Department of Biomedical Engineering, Lerner Research Institute, Cleveland Clinic, Cleveland, OH, USA

V. Abbaraju
Department of Electrical, Computer, and Systems Engineering, Case Western Reserve University, Cleveland, OH, USA

J. Han
Department of Biomedical Engineering, Lerner Research Institute, Cleveland Clinic, Cleveland, OH, USA

Department of Biomedical Engineering, Case Western Reserve University, Cleveland, OH, USA

L. Brody
SRS Medical, North Billerica, MA, USA

M. Damaser
Advanced Platform Technology Center, Veterans Affairs Medical Center, Cleveland, OH, USA

Department of Biomedical Engineering, Lerner Research Institute, Cleveland Clinic, Cleveland, OH, USA

A. Ahmed, J. Picone (eds.), *Machine Learning Applications in Medicine and Biology*,
https://doi.org/10.1007/978-3-031-51893-5_4

with spinal cord injury (SCI) are at increased risk for neurogenic lower urinary tract dysfunction (NLUTD), which is caused by a disturbance in the coordination between the central nervous system and somatic nervous system and leads to impaired control of the bladder [5]. LUTD diagnosis is multifaceted, but a commonly performed, quantitative assessment is the urodynamics (UDS) study.

The objective of UDS is to characterize the relaxation and contractile behavior of the bladder by monitoring its pressure and the volume as it fills and empties. The primary concept behind UDS is that it reproduces symptoms of LUTD quantitatively, enabling more precise diagnosis and treatment planning. While this is in principle true, in practice UDS suffers from many drawbacks and often fails to reproduce patient symptoms [6–8]. In addition, the test is discomforting and distressing to the individual [9].

One aspect of UDS is filling cystometry, retrograde filling of the bladder with saline or water to assess lower urinary tract function and dysfunction [1, 5]. Cystometry takes place in a clinical exam room and requires the insertion of two catheters: one intraurethral catheter for measuring vesical pressure (P_{VES}), i.e., pressure within the bladder, and one vaginal or rectal catheter for measuring abdominal pressure (P_{ABD}), i.e., pressure surrounding the bladder [5, 10]. This arrangement is depicted in Fig. 1. The catheters are either water or air filled to couple the pressure to an external transducer. The detrusor muscle pressure (P_{DET}) is the component of vesical pressure created solely by passive and active forces in the bladder wall; it is estimated by the simultaneous difference between vesical and abdominal pressure:

$$P_{DET} = P_{VES} - P_{ABD} \tag{1}$$

Both the P_{ABD} and P_{DET} signals are analyzed subjectively by urologists after UDS to diagnose different types of incontinence, such as stress urinary incontinence, urge incontinence, or mixed incontinence [1]. The P_{VES} signal alone is generally not evaluated due to the superimposition of P_{ABD}.

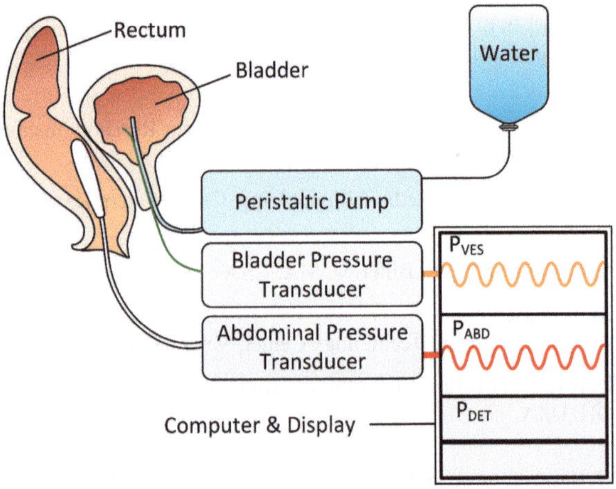

Fig. 1 Urodynamic cystometry uses a catheter placed in the bladder and rectum. This two-catheter approach allows estimation of the pressure generated by the bladder (detrusor) muscle, P_{DET}, from vesical (P_{VES}) and abdominal (P_{ABD}) pressure measurements

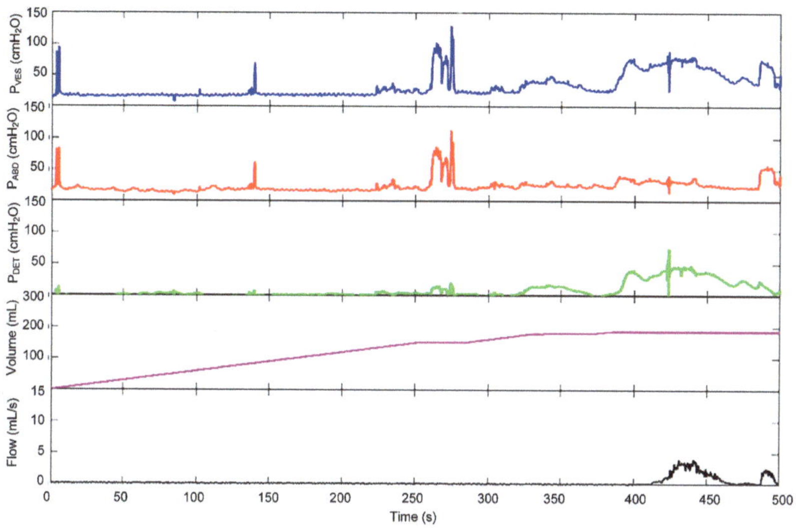

Fig. 2 Urodynamics recordings of vesical pressure (P_{VES}), abdominal pressure (P_{ABD}), calculated detrusor pressure (P_{DET}), infused bladder volume, and expelled urine flow rate are traditionally analyzed by a urology care team. Collecting these data requires the patient to sit still while catheters are inserted and the test is performed. The motivation in this work—catheter-free ambulatory urodynamics—is to collect the P_{VES} signal and extract the P_{ABD} and P_{DET} signals for interpretation

During filling cystometry, the patient is also asked to perform a series of abdominal maneuvers (e.g., cough and Valsalva) to ensure that the P_{ABD} and P_{VES} signals are responsive such that the P_{DET} signal can be accurately calculated [10]. Cystometry ends with the patient voiding (i.e., emptying) their bladder. Along with pressure measurements, the volume of the bladder and the flow rate of any voided liquid are measured and displayed in real time to monitor any leaks before the final voiding procedure. These five measurements (along with annotations of patient feelings of fullness, pain, urgency, etc.) form a UDS trace, from which a urologist observes and quantifies several key bladder events (Fig. 2). Analyzing the UDS result helps urologists determine a treatment plan for conditions like overactive bladder or neurogenic urinary incontinence [5, 10].

1.1 Single-Channel and Ambulatory Urodynamics

Traditional UDS has noted shortcomings relating to both the discomfort of the procedure, the artificial method of filling the bladder externally, and the limited duration (typically 1 hour) for observation of bladder behavior [9]. Ambulatory UDS—in which catheters are inserted like traditional UDS, but connected to a mobile data logging system—allows longer observation of bladder activity during natural filling. However, ambulatory UDS has not achieved significant adoption, in part due to the need for two catheters, over even longer durations than conventional UDS [6, 11]. Additionally, ambulatory UDS produces longer recordings of bladder behavior which are not annotated and must therefore be manually examined and annotated.

Recent efforts have focused on improving both catheter-based UDS and ambulatory UDS using only a single-channel measuring P_{VES}. This can be done through either a catheter or a wireless sensor [11]. A single-sensor approach would remove the P_{ABD} measurement entirely as well as the approximation for P_{DET} (Eq. 1). However, because the P_{ABD} signal is superimposed upon the P_{VES} signal, this suggests that measurement of P_{VES} alone is sufficient to recreate the traditional three UDS traces (P_{VES}, P_{ABD}, and P_{DET}), assuming the P_{ABD} signal can be extracted accurately. The extraction of this signal is the focus of this chapter, as described below.

1.2 UDS Interpretation

Although pressure is measured in UDS, it is not the only signal of interest for clinical interpretation. Besides the patient's sensations (which cannot be easily sensed and decoded except by the patient), bladder volume and urine flow rate are also recorded in traditional filling cystometry. Bladder volume is effectively known at all times, because it is controlled by the volume of saline infused by the external pump, and flow rate (and volume) during voiding or leaks is measured by capturing and weighing any expelled liquid.

UDS interpretation is both subjective and quantitative. Some measurements, such as bladder compliance and voiding pressure at maximum flow, are directly measured and used along with clinical guidelines for treatment decision-making. Because they require accurate measures of bladder volume, these measures cannot always be attained with pressure measurements alone (as in ambulatory systems). However, these quantitative measures still require a measurement of detrusor pressure. Additionally, subjective and semi-quantitative interpretation of UDS pressures is used to determine the presence of some urinary events and the bladder volume or leakage associated with them.

The annotation and interpretation of UDS traces, however, is subjective and variable. This is in part due to technician-related errors, motion artifacts during cystometry that may be falsely interpreted as events, or mispositioning/movement of the abdominal pressure catheter during the exam [12]. Furthermore, there exists a lack of clinical standardization in UDS protocol and interpretation from site to site [13]. Ongoing research efforts are focused on different algorithmic approaches for standardized UDS annotation through automated bladder event detection and classification [14].

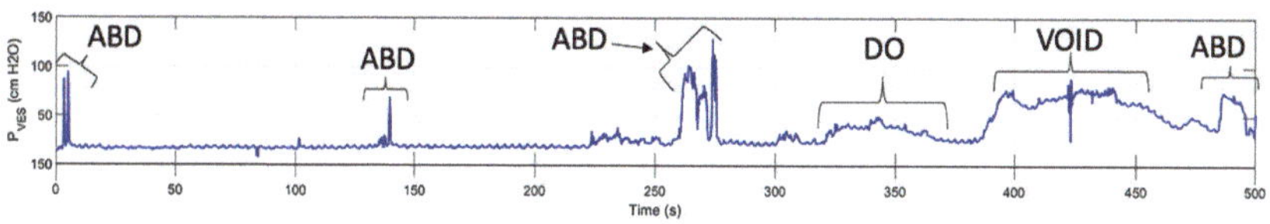

Fig. 3 Isolated P_{VES} trace after annotating urological events, as used for signal processing algorithm development. Note that the P_{VES} trace contains both abdominal (ABD) pressure events and those generated by the detrusor (VOID and DO). This property suggests that these signals may be independently extracted from the P_{VES} trace

This chapter partly focuses on the reproduction and automated detection of events during UDS, because the presence (or absence) of events is used in UDS interpretation [1, 5]. In this work we specifically examine the following events:

1. An abdominal event (ABD) generated by the abdominal muscles or bowel and superimposed on detrusor pressure when measured on P_{VES}. While P_{ABD} can arise from several causes, here we sub-divide ABD events into two classes: cough and Valsalva. Pressures caused by coughs (or laughs) are captured as a high-frequency spike in both P_{VES} and P_{ABD}. The Valsalva maneuver is a "bearing-down" motion which is an intentional, prolonged lower abdomen contraction used by some individuals to aid in urinary voiding.
2. Detrusor overactivity (DO), an involuntary detrusor contraction that occurs during the filling phase, is typically represented as a rise and fall in P_{VES} with no accompanying rise in abdominal pressure; furthermore, DO may result in leakage (i.e., incontinence), which is captured by a short period of slow, nonzero flow rate
3. A voiding contraction (VOID), a bladder contraction that occurs during the voiding phase and signifies emptying of the bladder by the patient, is a rise and fall in vesical pressure with a fast flow rate [4] (Fig. 3).

1.3 Signal Processing to Enable Single-Channel UDS

Single-channel UDS has several potential benefits compared to conventional two-catheter UDS. Most obviously, a single-channel system uses half the catheters, reducing disposable costs. Additionally, removing the rectal catheter improves patient comfort and reduces the likelihood of P_{ABD} artifacts caused by improper placement or shifting position of this catheter. Additionally, recent work on wireless bladder pressure sensors [11, 15–17] suggests that ambulatory recordings of P_{VES} may soon be available and require signal processing to extract the P_{ABD} signal for an accurate measurement of P_{DET}.

This chapter reflects two new signal processing approaches for managing single-channel UDS, whether from a catheter system, or an ambulatory wireless device. The chapter is organized in three parts, beginning with an overview of recent related work. Next, an adaptive single-channel UDS algorithm is presented, along with test results from data recorded from human patients. The adaptive UDS algorithm seeks to extract P_{ABD} accurately, using a feature threshold-based detection of specific common abdominal pressure signals. Finally, we present a classifier for detecting common abdominal and bladder pressure events. This approach would enable real-time detection of bladder events either for automatic annotation, data compression, or to improve the accuracy of P_{DET} measurement using only a measurement of P_{VES}.

2 Prior Work in Bladder Pressure Signal Processing

This section provides an overview of prior research in automated urological event classification from UDS data. Artificial intelligence and detection algorithms to improve decision-making are potentially transformative shift within urology [18]. Prior work on extracting frequency-domain information from UDS data [19, 20] suggested that patients may be stratified into finer categories of LUTD or that automated interpretation of DO may be feasible. More recent approaches leveraging signal processing and machine learning are reviewed in this section. The framework presented in this chapter is built on these approaches by applying ML with the additional constraints of a single-channel input and real-time operation.

The first paper focused on an algorithm for detecting bladder contractions from single-channel vesical pressure data and was designed for real-time implementation using energy-efficient signal processing techniques [21]. The last two employed learning-based approaches for DO event detection from multi-channel UDS data [14, 22]; these were the first published research using machine learning (ML) for urological event detection.

2.1 Prior Work: Context-Aware Thresholding for Bladder Contraction Detection

The context-aware thresholding (CAT) algorithm developed by Karam et al. was the first published approach for automated urological event detection from single-channel vesical pressure recordings. Developed for use in applications such as closed-loop bladder neuromodulation and data compression, this algorithm used a wavelet-based adaptive thresholding scheme to detect bladder contractions in real time [21]. CAT began by preprocessing the input P_{VES} signal using a low-pass exponential moving average filter (EMA) with a tunable cutoff frequency to remove high-frequency abdominal pressures caused by coughs and laughs. The five-level discrete wavelet transform (DWT) was then applied on the filtered signal to obtain approximation coefficients and detail coefficients.

Following the extraction of approximation and detail coefficients, adaptive thresholding was applied for event detection. First, a new approximation coefficient threshold was calculated from a buffer of stored coefficient values based on a percentile chosen to minimize false positives while allowing true-positive detection. Then, if the most recently calculated approximation coefficient exceeded this threshold, a bladder contraction is detected. This approximation coefficient was then stored in the buffer for calculating a new threshold to process the next P_{VES} sample. A similar adaptive thresholding scheme was applied on the detail coefficients to detect abdominal/motion artifacts in real time, as artifacts were represented by high-frequency spikes in pressure captured in detail coefficients. The CAT algorithm was tested on pre-recorded UDS data and demonstrated 97% accuracy in detecting bladder voiding contractions within 1 second of onset. Application of the CAT algorithm to trigger electrical stimulation from wireless pressure sensor data to treat urinary incontinence in live animals has since been demonstrated [15].

A key advantage of the CAT algorithm was flexibility. Since the algorithm did not employ any artificial intelligence or ML, its tunable parameters—like the cutoff frequency of the EMA filter and the percentile value for computing the adaptive coefficient thresholds—could be easily adjusted prior to program startup without having to "re-train" the algorithm. Furthermore, CAT was suitable for low-power hardware implementation: the EMA filter was comprised of a few simple computational blocks, and the wavelet lifting scheme provided an energy-efficient version of the DWT [23].

The CAT algorithm thus provided real-time bladder contraction detection from single-channel P_{VES} data, which was demonstrated in both closed-loop bladder neuromodulation and urodynamic data compression. However, CAT only identified contractions and artifacts; it did not distinguish voiding contractions from non-voiding contractions (i.e., DO) or motion artifacts from abdominal events (i.e., cough or Valsalva).

2.2 *Prior Work: Manifold Learning for DO Detection*

For ambulatory urodynamics, which require differentiation to provide context for clinical analysis, binary outputs are not sufficient. To this end, ML approaches have recently been proposed to classify urological events with more specificity. The pattern recognition algorithm developed by Wang et al. was the first publication applying ML for urological event detection from UDS data [22]. Specifically, this framework was designed to automatically detect DO events from detrusor pressure using manifold learning, which is a branch of traditional ML. The core mechanism behind the proposed event detection framework is comparing a test P_{DET} segment to a *DO library*. The measured similarity between the test segment and the library indicated the presence or absence of DO in that segment [22].

Each UDS trace in the dataset, which included nearly 800 annotated UDS traces, was first pre-processed using a sixth-order high-pass Butterworth filter followed by a median filter to reduce noise and "flatten" the P_{DET} signal. Each trace was then windowed into 6-second segments based on the presence of DO labels, with each interval centered around the peak of the annotated DO event. Each interval, or "wave-shape," was stored as a separate vector, and this set of DO wave-shapes was parameterized by the wave-shape manifold model. This model mapped each vector onto a manifold, which was a high-dimensional, nonlinear surface from which a set of multivariate parameters were derived to define each wave-shape in terms of a generalized curve. For instance, a bell curve with multiple parameters describing its height, width, and curvature could serve as the wave-shape manifold for DO events.

K-means clustering (with $k = 5$) was then performed on the resulting bank of aggregated, parameterized DO wave-shapes to identify five basic DO patterns—i.e., the DO library. This library served as the basis of the DO detection algorithm presented by Wang et al. Using dynamic time warping, a method that measures the similarity between two sequences by stretching and contracting the input, the algorithm compared an input 6-second P_{DET} segment with each of the five basic DO patterns in the library. This produced a similarity score—with a higher score indicating a stronger likelihood of the input P_{DET} segment containing DO.

Training determined the optimal similarity threshold that indicated the presence of DO in a P_{det} segment. Since most 6-second segments in the dataset did not contain DO, 2% of all non-DO segments were randomly sampled and added to the dataset to maintain balance during training. Fivefold cross-validation evaluated the performance of the manifold learning DO detection algorithm, yielding accuracy, sensitivity and specificity scores of 81.4%, 76.9%, and 81.4%, respectively.

Identifying DO visually—let alone algorithmically—can be challenging due to the variety of irregular patterns/forms that the condition can take. The pattern recognition algorithm demonstrated by Wang et al. was thus impressive due to its ability to automate this process through manifold modeling and dynamic time warping. Furthermore, this algorithm may be suitable for real-time inference as it only requires 6 seconds of P_{DET} data to make a new prediction. However, a key disadvantage of this algorithm was that it was dependent on P_{DET}, which—in a typical UDS setting—requires the measurement of both P_{VES} and P_{ABD} using two catheters. Moreover, this algorithm only detected one type of urological event (DO), limiting its application in real-time ambulatory urodynamic annotation/analysis.

2.3 *Prior Work: Machine Learning with Time/Frequency Features for DO Detection*

Hobbs et al. further explored the application of post-hoc ML for DO event detection by experimenting with different feature extraction strategies to pair with a support vector machine classifier. Time- and frequency-based features were derived from either a single pressure channel (P_{VES}) or all three pressure channels (P_{VES}, P_{ABD}, and P_{DET}) [14].

The dataset consisted of around 800 pre-recorded and fully annotated UDS traces. Unlike previous research, no data pre-processing via filtering or smoothing was performed prior to training/testing. Each trace was windowed into 60-second segments, with successive segments overlapping by 20 seconds. For time domain feature extraction, the mean, minimum, maximum, median, and standard deviation of each segment were computed. In the case of time-based feature extraction using all three pressure channels, features from each channel were concatenated for each segment. For frequency domain feature extraction, a fast Fourier transform was applied on each 60-second segment. No additional features were computed; instead, principal component analysis (PCA) was performed to reduce feature dimensionality by choosing only 95% of the frequencies that contributed to the signal of interest. In the case of frequency-based feature extraction using all three pressure channels, the relevant frequencies from each signal were concatenated for each segment. Finally, the feature vector for 60-second segment was passed through the support vector machine classifier with the radial basis function kernel (RBF-SVM). In total, Hobbs et al. tested four different window-based feature extraction approaches:

1. Time domain features from P_{VES} alone
2. Time domain features from P_{VES}, P_{ABD} and P_{DET}
3. Frequency domain features from P_{VES} alone
4. Frequency domain features from P_{VES}, P_{ABD} and P_{DET}

Threefold cross-validation was performed to evaluate each of the different feature extraction strategies with the RBF-SVM classifier, which showed that using single-channel, time-based features resulted in the highest sensitivity (85.4%), while using three-channel, frequency-based features resulted in the highest specificity (92.9%). Hobbs et al. noted that predicting false DO positives most commonly occurred with either motion artifacts or abdominal events; including all three pressure channels reduced this false positive rate (thereby increasing specificity) but increased the false-negative rate (thereby decreasing sensitivity) [14].

Hobbs et al. thus presented the first study on how time- and frequency-domain features can serve as inputs for widely used ML classifiers to automatically detect DO [14]. A significant contribution of this work was its demonstration of feature extraction from single-channel vesical pressure data for the ML classification pipeline. However, much like the work by Wang et al. [22], the presented framework only detected a single type of urological event (DO). Furthermore, the use of long windows (60 seconds) for classification limited the potential for real-time inference and action (i.e., notifying the patient to use the restroom or applying neuromodulation to inhibit DO before leakage occurs).

3 Algorithmic Real-Time Single-Channel Detrusor Pressure Estimation

In single-channel urodynamics, only P_{VES} is measured since this the signal contains both the P_{ABD} and P_{DET}. Both signals are of interest in a urodynamics assessment—and urodynamics requires real-time pressure display for debugging and troubleshooting during the examination. Therefore, in this section we demonstrate two new approaches to algorithmically extract the unique detrusor and abdominal pressure contributions from the P_{VES} signal, with minimal delay. The framework for this algorithm relies partly on spectral differences between P_{ABD} and P_{DET} but also addresses times of spectral overlap with an adaptive event detection block.

The detrusor pressure (P_{DET}) is generated by the smooth muscle in the bladder, which has limitations on the speed and strength of muscle contractions [24]. Previous studies on single-channel urodynamic recordings have shown that wavelet-based algorithms can accurately identify meaningful bladder events using only the pressure within the bladder (P_{VES}) with a low rate of false positives [12, 13]. However, this approach requires a well-defined frequency boundary between P_{DET} and P_{ABD}, which is not necessarily guaranteed in practice. Furthermore, the accuracy of this approach is dependent on the specific wavelet used to ensure accurate separation of the P_{DET} and P_{ABD} signals. Additionally, even with an optimal wavelet choice, the boundary between P_{DET} and P_{ABD} may change over time due to changes in bladder pressure and posture.

3.1 *Urodynamic Data Test Set*

A total of 20 (7 male and 13 female) anonymized UDS recordings were collected using air-charged catheters with a CT3000Plus Complete Urodynamics system (SRS Medical, N. Billerica, MA) at a sampling rate of 10 Hz. Recordings consisted of multiple signals including volume voided (volume), intravesical pressure (P_{VES}), and intra-abdominal pressure as measured via a rectal catheter (P_{ABD}). A signal representing the simultaneous difference between these two pressure channels ($P_{DET} = P_{VES} - P_{ABD}$) was calculated. Data were manually annotated to denote relevant events such as cough, Valsalva, position change, and voiding contraction during retrograde bladder filling and voiding.

3.2 *Single-Channel Detrusor Pressure Estimation Framework*

The results of correlation analysis on 20 UDS recordings showed a linear relationship between P_{DET} and P_{VES}, suggesting that P_{DET} can be found within P_{VES}

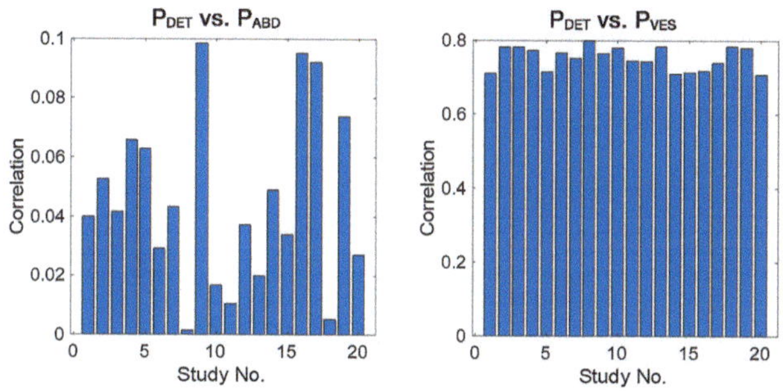

Fig. 4 Correlation between P_{ABD} and P_{DET} (left) as well as P_{VES} and P_{DET} (right) from data from 20 human subjects. The low correlation between P_{DET} and P_{ABD} indicated that P_{DET} is best extracted from the P_{VES} signal only

(Fig. 4). Furthermore, the low correlation between P_{DET} and P_{ABD} demonstrates that P_{ABD} is unrelated to P_{DET} and must be removed from the calculation of P_{DET}. As a result, the proposed framework uses P_{VES} to estimate P_{DET}. By extension, all pressures in P_{VES} not caused by P_{DET} are assigned to P_{ABD} via subtraction.

The framework consists of three main components: a prefiltering stage, a discrete wavelet multiresolution decomposer, and a signal reconstruction stage (as depicted in Fig. 5a). The framework has a straightforward structure, making it easier to verify and debug in an embedded system implementation, where P_{VES} is first stored in a buffer and then processed in the pipeline using a sliding window technique. The tuning of each stage is dependent on the size of the window, making the choice of sliding window size (W_{sz}) a factor in determining the efficiency of the estimation. The value for W_{sz} was selected after examining the least mean square reconstruction of P_{VES} from wavelet coefficients, as described in Sect. 3.3. Optimization of each stage in the pipeline was performed after choosing a window size W_{sz} of 32 samples. This is equivalent to 3.2 seconds of data history at the current standard P_{VES} sample rate of 10 Hz.

3.3 *Prefiltering and Discrete Wavelet Transform Stages*

The prefilter removes white noise and spike artifacts which arise from movement of catheters during cystometry. A two-stage filter was used. The first stage was a tenth-order finite impulse response band-pass filter with bandwidth of 0.02–5 Hz and stop band attenuation of 30 dB. The second stage was a third-order Savitzky-Golay smoother which used convolutional linear least squares to fit successive adjacent data points with a low-degree polynomial [25, 26]. Unlike simpler smoothing techniques, such as a moving average filter, the Savitzky-Golay smoother preserves data features such as peak width and height.

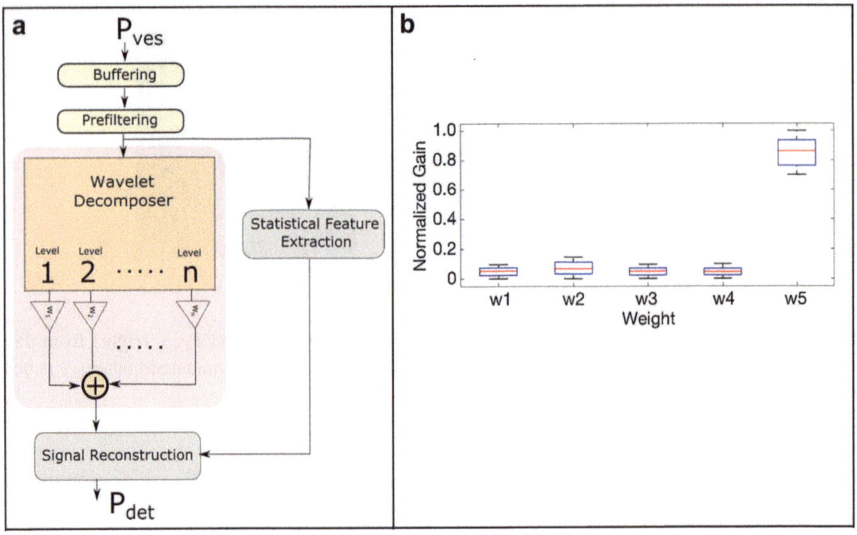

Fig. 5 (**a**) A block diagram shows the estimation process stages: filtration, wavelet decomposition, and signal reconstruction using the statistical features. (**b**) Gains in wavelet decomposition of 5-level wavelets resulting from LMS for UDS dataset and 32 window size

The filtered data were passed to a wavelet multiresolution analyzer (MRA) and decomposed into fundamental frequency components. The strength of MRA dwells in its ability to preserve time-domain and frequency-domain information using the DWT. The proposed MRA unit was augmented with a tunable weight vector ω_i, where $i \in [1, 2, \ldots, M]$ and $M = \log_2(W_z)$ was the wavelet resolution level. The weight vector provided a flexible method of estimating P_{DET} using a weighted sum. With $W_{\mathrm{sz}} = 32$, the 32-sample frame decomposed into five frames representing the window component at 1, 2, 3, 5, and 10 Hz. Due to decimation at each stage of the transform, the final level of the transform yielded only a single approximation coefficient.

In addition to window size, sampling frequency, and weighting vector, selecting a mother wavelet for MRA played an important role in estimating P_{DET}. The output of the wavelet decomposer, $\overline{P}$, can be described as:

$$\overline{P}\,(W, \psi, W_{\mathrm{sz}}) = \sum\nolimits_{i=1}^{M} \left(\frac{1}{\sqrt{2}}\right)^{i} W_i \psi_i \tag{2}$$

where W_i is the window resolution at level i, ψ (.) is the mother wavelet function, and M is the depth of decomposition. Often, the vector W_{sz} is predefined and selected to capture the slowest event of interest. Instead, we investigated varying the mother wavelet first with a unity weighted vector value. Then an appropriate W_{sz} value was obtained with a static mother wavelet function via least mean square (LMS) fitting.

Table 1 The characteristic of reconstruction outcomes using different wavelet functions

	Haar	DB4	Sym4	Dmey	Coif4
Reconstruction computation time (RT) (ms)	1.9	3.7	**3.4**	8.7	14
Reconstruction error (RE)	1e-5	1e-12	**1e-12**	1e-12	1e-12

Reconstructing a 10 minutes cystometry signal from its MRA resolutions provided a test for wavelet function fidelity. The root mean square (RMS) error between the original and reconstructed signal—called reconstruction error (RE)– and the reconstruction computation time (RT) were calculated. The RT can be defined as the time needed to complete reconstructing 1000 samples of data, which mainly depends on the hardware capability. Here, RT is expressed in milliseconds when comparing results obtained from a laptop running Matlab. RT and RE were used to evaluate the selected wavelet function though the reconstruction test. A set of popular symmetrical and orthogonal wavelet functions were selected to test as shown in Table 1. The Symlet wave function with four vanishing points (Sym4) was selected as it achieved the best balance between RT and RE [27–29].

To find the weight vector, LMS fitting was used with P_{VES} as the input and P_{DET} as the output. With $W_{\mathrm{sz}} = 32$ and wavelet function Sym4, the P_{VES} was decomposed into five levels. The LMS results showed that the last level weight was significantly higher than the other weights when estimating the low-frequency P_{DET} signal. Figure 5b shows the average and variation of weights after applying the LMS optimization to the 20 UDS studies. This result was consistent with previous UDS analyses which used Daubechies wavelets [21].

3.4 Event Detection and Elimination Stage

Event detection improves the DWT estimate of P_{DET} because it allows an event-dependent selection of the reconstruction weights W_i. For instance, if an abdominal event (containing high-frequency elements) is detected, the weighting vector for windows during the event can de-emphasize initial scales of the DWT. The reconstruction weights W_i are also adaptively calculated and can therefore capture the recent history of activity in the data, enabling a more robust detection scheme. In our approach, we focused on statistical features for adaptively scaling W_i. Statistical features of the window serve as an alternative abstract representation of the window and a strong indicator of contraction onset and termination of each event.

However, while previous studies reported only detection of event onset, detecting the total span of an event is an important factor that has received less attention [7]. The span of the event is the time difference between the onset and termination time. Accurately capturing the event span is important in single-channel UDS, since it enables the signal processing algorithm to revert to the more accurate P_{DET} estimation via DWT approximation coefficients after an event ends. The given UDS dataset showed that the nominal duration of coughs and Valsalva maneuver was 1.4 ± 0.5 seconds and 2.5 ± 0.5 seconds respectively. The fixed window of 32 samples captured the longest Valsalva events with at most two consecutive windows. An event ending flag was added to the event detection kernel to allow detection of events spanning more than one window.

Event detection used the previous observations to identify a vector of threshold values to identify the event onset and termination. For a sliding window, the kernel function of the event detection algorithm was:

$$\Gamma(k) = \begin{cases} \mathbf{1} & F_g > T_m \text{ and } f(v) \\ \mathbf{0} & \text{otherwise} \end{cases} \tag{3}$$

where $v = [\mu, \sigma, \xi, \delta] \in \mathbb{R}^4$ is a vector, k is the discrete time index, $\Gamma(k)$ is a discrete logical function that signifies event detection, yielding a true value when events occur, σ is the local window standard deviation, μ is the arithmetic mean, δ is the maximum value of local gradient, ξ is the rate of signal zero crossing, F_g is a flag set at onset and cleared at termination, and T_m is a tuning parameter. Function f is the core of the kernel function which is a set of inequalities that they are logically true only during events. This function is simply selected as $f = \text{AND}(v > \mathbf{T_r})$, where $\mathbf{T_r} \in \mathbb{R}^4$ is a vector of threshold values. In general, the estimated P_{DET} was calculated by eliminating the detected events, i.e.:

$$P_{DET}(W, W_{sz}, \mu, \sigma, \delta, \xi) = \begin{cases} \overline{P} & \Gamma(k) = 1 \\ \overline{P} - \mu & \text{otherwise} \end{cases} \tag{4}$$

This relatively simple technique was adopted due to the low computational overhead, which makes it more suitable for real-time implementation, e.g., on portable UDS equipment.

3.5 Algorithm Performance and Accuracy

In addition to smoothing the P_{VES} signal, the two-stage prefilter attenuated large amplitude, pressure spikes due to coughs (Fig. 6). Coughs should only appear in P_{ABD}, with no pressure in the P_{DET} estimation, because they are generated entirely by the abdominal muscles. Pairing the wavelet decomposer with the prefilter resulted in an average 23% amplitude reduction in cough artifacts passing to the P_{DET} estimation instead of P_{ABD}. Further attenuation of coughs via a lower filtering cutoff was not feasible beyond this level, due to attenuation of actual P_{DET} transients.

The RMS value and correlation coefficient R were used to evaluate algorithm accuracy. Furthermore, a statistical F-score was used to evaluate the accuracy of artifact detection. The output of the estimator was compared with the calculated

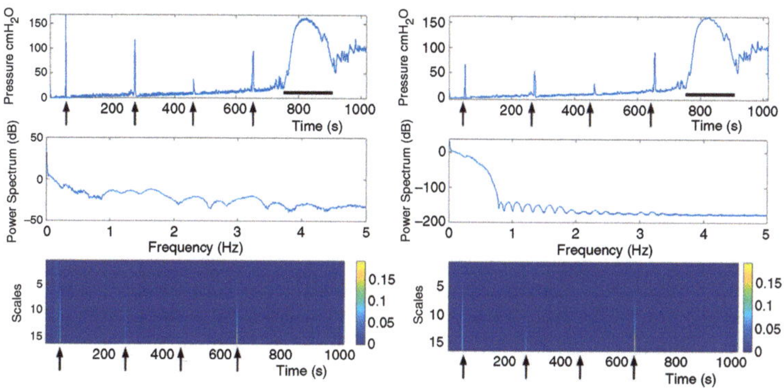

Fig. 6 Frequency spectrum and scalogram of P_{VES} before (left) and after (right) the two stage prefilter. Arrows indicate the locations of cough events, and the solid horizontal bar identifies the duration of the bladder voiding (emptying) phase

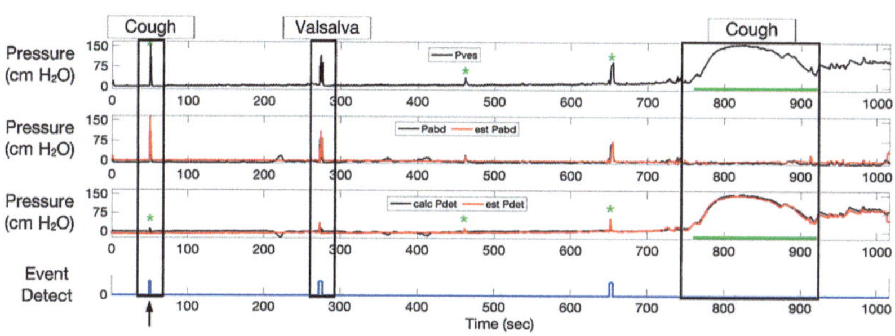

Fig. 7 A urodynamic study with measured P_{VES} (first row), measured P_{ABD} and estimated P_{ABD} (second row), estimated P_{DET} and calculated P_{DET} (third row), and event indicator (fourth row). Cough events on P_{VES} and passing to P_{DET} are marked with asterisk (*) and the void event is denoted by a solid horizontal line indicating the start and end

P_{DET}, and the overall estimation performance showed RMS $= 10.7 \pm 2.1$ cmH_2O and $R = 0.88 \pm 0.6$ ($N = 20$). Moreover, detection accuracy for cough and Valsalva were 99.5% and 84.3%, respectively (Fig. 7). Detection accuracy was determined relative to clinical annotation of events for each cystometry recording.

Although this approach effectively removed identified classes of abdominal events from corrupting the output P_{DET} trace, abdominal events which mimicked bladder contractions were not removed (Fig. 8). Because these events are very long with significant low-frequency activity, neither the DWT estimator nor event detector algorithm can differentiate them from P_{DET} contractions. While retrospective analysis of UDS waveforms allows detection of these events (because the entire event duration is apparent), a real-time implementation cannot introduce significant delay. Therefore, further work, perhaps in the form of machine learning classifiers (e.g., those with long- to short-term memory structures) may be needed in the future.

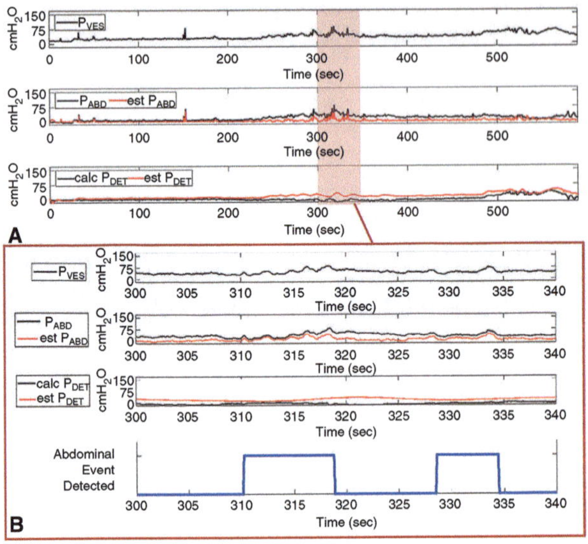

Fig. 8 Example processing of a UDS P_{VES} trace via the single-channel estimator (**a**). This example is shown to demonstrate shortcomings of the proposed event detection methodology, which is limited to relatively short windows. Long abdominal events are not continuously detected, to avoid inaccuracy when estimating the P_{DET} signal, which is essential for treatment planning (**b**)

4 Machine Learning Classification of Urinary Events from Single-Catheter Data

The previous section demonstrated an algorithm for real-time extraction of the P_{DET} and P_{ABD} signals from a single P_{VES} measurement channel. Nominally, reconstruction of P_{DET} used a weighted sum of DWT approximation coefficients, while several statistical features were compared against thresholds to adaptively modify the reconstruction weights based on the presence of known abdominal events. The detection of these events, however, may be improved using supervised machine learning (ML). In this section, we focus on the classification of events to improve the accuracy of single-channel urodynamics. In this phase of research, classification was not performed in real time; since the presented approaches were developed using features which may be derived causally, ongoing research efforts are focused on transforming these approaches for real-time operation. Specifically, these approaches are developed using features which may be derived causally. Real-time implementation of these ML techniques is broadly applicable to other applications involving bladder pressure data processing. Two examples not discussed here include on-the-fly data compression during ambulatory urodynamics [11] and automatic detection of bladder contractions to improve neuromodulation [15, 30].

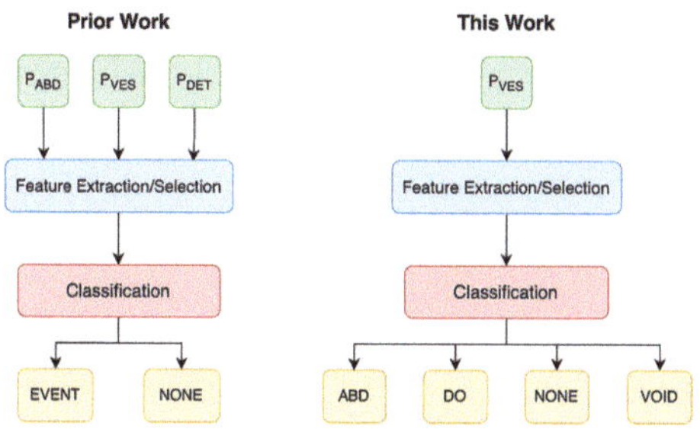

Fig. 9 Comparison of classification schemes posed by prior work (left) and the work presented in this here. This work applies multi-event classification from a single-channel input (P_{VES}). The general framework is motivated by single-catheter urodynamics and emerging catheter-free urodynamics applications

Therefore, this section describes the post-hoc ML framework design process for automated, single-sensor urological event classification. The design began with preparing an annotated dataset with the assistance of a physician and subsequently splitting the dataset into uniform-length, non-overlapping segments. Then, a feature extraction pipeline was designed by employing the DWT from which various statistical values were computed. Finally, three different classifiers were chosen to train and test on the dataset, with each classifier using a different number of features determined using ReliefF-based feature selection. While prior work in this domain focused on single event detection and in some cases from multiple pressure channels, this approach demonstrated classification of three types of events from a single-channel input: P_{VES} (Fig. 9).

4.1 *Dataset Preparation for Classifier Training*

A total of 108 UDS traces were obtained from human patients with overactive bladder or neurogenic urinary incontinence. All data were anonymized and harvested from previous studies, which were IRB exempt. Urodynamic testing for these studies was conducted at the Louis Stokes Cleveland Department of Veteran Affairs Medical Center and the Cleveland Clinic. Unlike the dataset analyzed in Sect. 3, these data were collected using water-filled catheters, and the datasets were not combined. Both air- and water-filled catheters are used for UDS testing clinically.

Each trace included P_{VES}, P_{ABD}, P_{DET}, bladder volume, and flow rate and, with the assistance of a blinded urologist, was annotated with three urological events: abdominal event (ABD), detrusor overactivity (DO) and voiding contraction (VOID) (previously shown in Fig. 3). The P_{VES} data of each trace along with the annotated labels were then isolated, with noisy segments caused by catheter insertion or removal—found near the start or end of a trace—discarded from each trace.

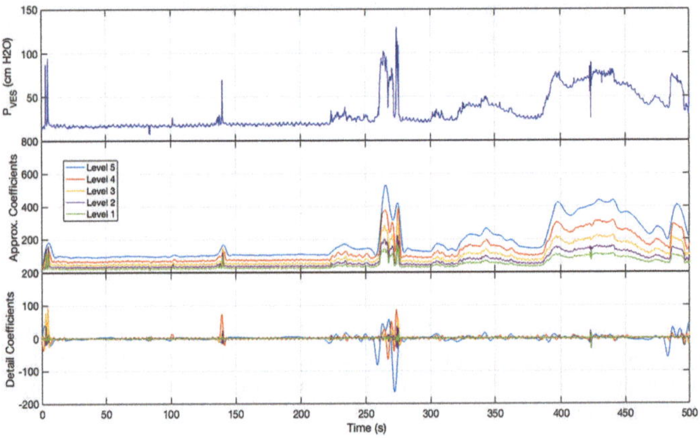

Fig. 10 Output of 5-level DWT applied to P_{VES} followed by cubic spline interpolation, resulting in 10 total signals: 5 sets of approx. coefficients and 5 sets of detail coefficients

4.2 Segmental Feature Extraction

Karam et al. demonstrated that using the Db2 wavelet with up to 5 DWT levels resulted in useful features for time-frequency analysis of urological events [21]. As such, the ML framework we used applied the five-level DWT with the Db2 wavelet over the full length of each P_{VES} trace in the dataset. Then, to account for the downsampling operations performed each DWT level and to keep the time scale uniform across all outputs for optimal classifier performance, cubic spline interpolation was applied to each set of coefficients (by a factor of 2^m, where m denotes the transform level (Fig. 10)). For a real-time implementation, the lifting wavelet transform algorithm can be used to reduce computational and memory overhead [21]. However, in real-time applications, the entire P_{VES} trace is *not* available, which reduces DWT resolution and may negatively impact event detection accuracy. Nonetheless, a fixed DWT window size D and shift length T can be chosen, and the DWT window can shift in real time to consider only the newest T samples and the previous $D - T$ samples for obtaining approximation and detail coefficients. Similarly, cubic spline interpolation may be expensive to implement on a low-power device; foregoing this step in a real-time implementation would present another compromise between improved inference latency and decreased detection accuracy.

After performing cubic spline interpolation, each set of approximation coefficients and detail coefficients contained N elements, where N was the length of the input P_{VES} signal. Every individual P_{VES} datapoint and its corresponding 10 DWT coefficients could have thus been treated as features for the classifier, which could train to predict the class of each individual datapoint. However, with a future real-time implementation in mind, where it is costly to compute features and make a new prediction for every incoming sample, the ML classifiers were trained to classify *segments* of collected P_{VES} data. Furthermore, training the classifier on data segments as opposed to individual samples allowed for more statistical features—e.g., the mean, median, maximum, etc. over a *range* of data—to be computed. Therefore, to balance feature computation range with minimal inference delay for future real-time implementation, 0.8 second was chosen as the segment length (8 P_{VES} samples at a 10 Hz sampling rate). Each P_{VES} trace in the dataset and its corresponding DWT outputs were thus split into non-overlapping 0.8 second segments, and each segment was labeled with one of four possible class labels, depending on which event was present in the segment: ABD, DO, VOID, or NONE (if no event was present).

Following data segmentation, features were computed based on the approximation and detail coefficients extracted from the five-level DWT. In prior work classifying time-varying physiological signals like ECG and EEG using the DWT in tandem with ML, statistical properties such as the mean, maximum, Shannon entropy, power, etc. from each DWT sub-band were used as features [28, 31, 32]. Adopting this approach for vesical pressure, for each 0.8-second P_{VES} segment, the maximum, mean absolute value (MAV), median, and Shannon entropy from each of the ten corresponding DWT output segments were computed, resulting in 40 total features for each P_{VES} segment. Furthermore, the cross-correlation signal between approximation coefficients and detail coefficients in each DWT level was computed, resulting in five cross-correlation vectors for each segment. From these vectors, the maximum, mean, and median were extracted, resulting in 15 additional features for each P_{VES} segment (Fig. 11). For a real-time implementation, where it may be expensive to compute the cross-correlation every 0.8 second, the ratio between approximation and detail coefficients in each level can be used as features that approximate the similarity between the signal's low-frequency and high-frequency content.

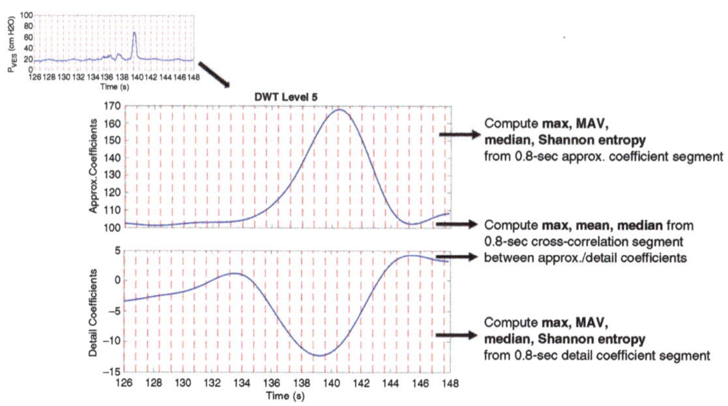

Fig. 11 Statistical feature extraction from the fifth level of the DWT after P_{VES} segmentation. This process was performed for all five DWT levels, resulting in 55 features for each 0.8-second P_{VES} segment

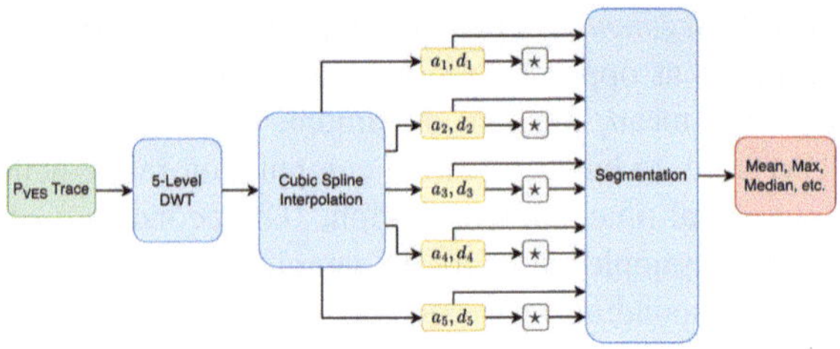

Fig. 12 Full feature extraction pipeline starting with P_{VES}. The ⋆ symbol represents the cross-correlation operator. A total of 55 features were computed for every individual 0.8-second P_{VES} segment

Table 2 List of features extracted from each DWT level after cubic spline interpolation

Signal	Features
Approximation coefficient segment ($N = 8$)	MaximumMean absolute valueMedianShannon entropy
Detail coefficient segment ($N = 8$)	MaximumMean absolute valueMedianShannon entropy
Cross-correlation segment ($N = 16$)	MaximumMeanMedian

Therefore, for each 0.8-second interval of P_{VES} data, a total of 55 features were extracted. 85,780 0.8-second P_{VES} segments from the entire dataset containing 108 UDS traces were combined to form an initial 85,780 × 55 feature matrix. Most of the dataset consisted of NONE segments, while the ABD class was the least represented. To maintain balance during classifier training, 4000 P_{VES} segments corresponding to each class were randomly chosen, forming a final 16,000 × 55 feature matrix for cross-validation. A summary of the full feature extraction pipeline is shown in Fig. 12, and the features extracted are listed in Table 2.

4.3 Feature Selection and Hyperparameter Tuning

Three initial classifier models for the ML framework were evaluated: (1) a k-nearest neighbor classifier (KNN), (2) an artificial feed-forward neural network classifier with two hidden layers (ANN), and (3) a support vector machine classifier with the radial basis function kernel (SVM-RBF). The number of ANN hidden layers and SVM kernel was chosen based on prior work in physiological signal classification using DWT_based features [28, 31]. Feature selection was applied to reduce the training time and overfitting of the classifiers [28, 31]. The most relevant features from the feature matrix were identified using the ReliefF algorithm, which assigns weights to features using a nearest-neighbor approach [33]. ReliefF rewards features with large differences for observations of different classes and penalizes features with large differences for observations of the same class [33]. This technique identified the m highest-ranked features, where m denotes the number of features chosen for each classifier architecture. Feature selection may also be useful for a real-time implementation of the presented ML methods as it would reduce the number of features that need to be computed and thus reduce the computational cost of the overall algorithm.

Fig. 13 Process for iterative feature selection and hyperparameter tuning for each classifier. The feature set was determined prior to tuning

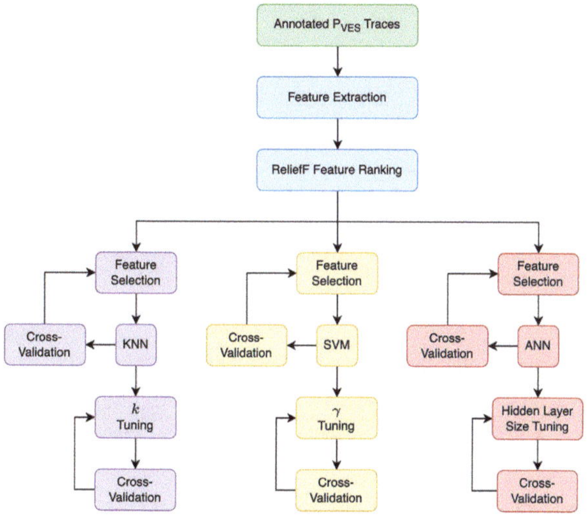

Table 3 Final classifier architectures for single-sensor urological event classification

Classifier	Hyperparameters	Number of features
k-Nearest neighbors (KNN)	$k = 1$	17
Support vector machine (SVM)	Kernel: radial basis function (RBF) $\gamma = 1.2$	39
Artificial neural network (ANN)	Hidden layers: 2 × 100 neurons/layer Activation function: rectified linear unit (ReLU)	45

Following feature selection, the hyperparameters for each model were tuned by evaluating the fivefold cross-validation accuracy of each model. These hyperparameters included the k value for the KNN, the number of hidden layer neurons for the ANN, and the γ value for the SVM-RBF. Cross-validation, described in more detail in Sect. 4.4, helped evaluate the generalizability of each model by accounting for potential sources of bias and variance. The process for feature selection and hyperparameter tuning is summarized in Fig. 13, and the final classifier architectures with number of features and hyperparameter values are shown in Table 3.

4.4 Classifier Validation on Pre-recorded Urodynamic Data

After data segmentation and random sampling (to achieve a balanced distribution from each class), the final dataset consisted of 16,000 0.8-second P_{ves} segments, with each segment belonging to one of four classes: ABD, DO, NONE, or VOID. To evaluate the performance of the three classifiers—each of which underwent both feature selection and hyperparameter tuning—fivefold cross-validation (CV) was

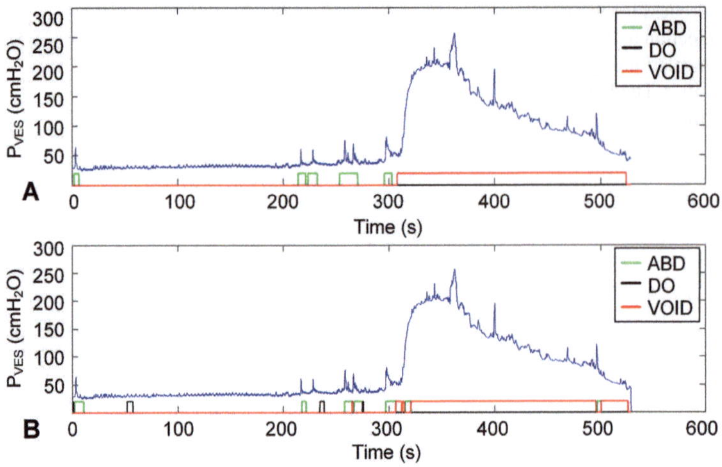

Fig. 14 Example time-domain view of classifier output. Urologist estimates indicated abdominal (ABD) activity preceding the voiding (VOID) event (**a**). Although the real-time classifier used independent 0.8-second segments of data, it generally classified the events correctly in this example (**b**)

performed (Fig. 14). In each fold, 80% of the dataset (12,800 P_{ves} segments) was used for training, and the overall CV accuracy was the average testing classification accuracy over all five folds. After CV, the following performance metrics were computed for each class where TP, FP, TN, and FN represent the number of true positives, false positives, true negatives, and false negatives, respectively [34]:

$$\text{Sensitivity (Recall)} = \frac{\text{TN}}{\text{TP} + \text{FP}} \tag{5}$$

$$\text{Specificity} = \frac{\text{TN}}{\text{TN} + \text{FP}} \tag{6}$$

$$\text{Precision} = \frac{\text{TP}}{\text{TP} + \text{FP}} \tag{7}$$

$$\text{F1} = \frac{2 \times (\text{Precision} \times \text{Recall})}{\text{Precision} + \text{Recall}} \tag{8}$$

Computing these performance metrics was advantageous as, depending on the application, each type of urological event suffers from misclassification differently. For instance, in real-time bladder neuromodulation, the classifier must have a high precision and a high specificity for VOID events to avoid unnecessarily stimulating the patient. For real-time data compression, the cost of a false positive for DO events

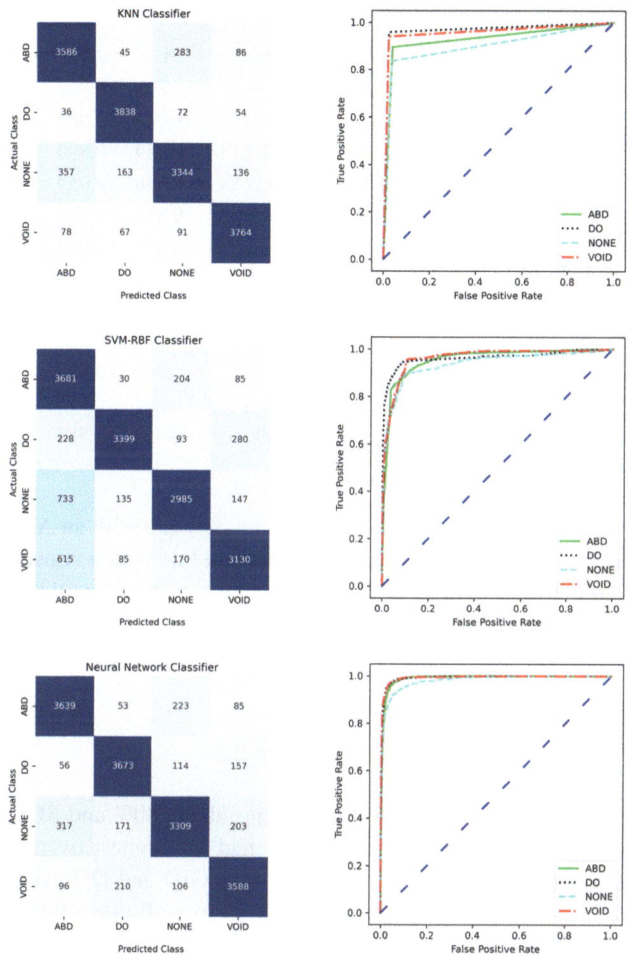

Fig. 15 Confusion matrix and ROC curve from fivefold cross validation testing for each classifier

is not as high, as over-annotating these events can inform a patient to undergo further screening, while under-annotating may result in a missed diagnosis of LUTD. In this case, high sensitivity to DO events could serve as a useful performance metric. For applications that require both high precision and recall (sensitivity), F1 score is suitable as it balances the two [35].

Finally, the receiver operating characteristic (ROC) curve and the area under the curve (AUC) per class were calculated for each classifier. The ROC curve plots the false-positive rate (FPR, i.e., inverted specificity) versus true-positive rate (TPR, i.e., sensitivity) as a function of various classification thresholds [36] and represents how a classifier treats different types of misclassifications. The AUC summarizes the skill of each classifier based on these ROC curves; the ROC of a perfect classifier

Table 4 Per-class performance metrics for each classifier

Classifier	Event	Specificity	Sensitivity	Precision	F1	AUC	Accuracy (%)
KNN	ABD	0.961	0.897	0.884	0.890	0.929	90.8
	DO	0.977	0.960	0.933	0.946	0.968	
	NONE	0.963	0.836	0.882	0.859	0.899	
	VOID	0.977	0.941	0.932	0.936	0.959	
SVM	ABD	0.869	0.920	0.700	0.795	0.962	82.5
	DO	0.979	0.848	0.931	0.888	0.964	
	NONE	0.961	0.746	0.865	0.801	0.944	
	VOID	0.957	0.782	0.859	0.819	0.959	
ANN	ABD	0.961	0.910	0.886	0.898	0.985	88.8
	DO	0.964	0.918	0.894	0.906	0.985	
	NONE	0.963	0.827	0.882	0.854	0.972	
	VOID	0.963	0.897	0.890	0.893	0.984	

would begin at a (FPR, TPR) of (0, 1) and remain constant with an AUC of 1, while the ROC of a "no-skill classifier" would appear as a line with a constant slope of 1 and an AUC of 0.5. The summary of results from CV testing for all three classifiers is shown in Fig. 15 and Table 4.

4.5 Classifier Performance Summary

All three classifiers scored overall CV accuracies above 80% and ROC AUCs at or above 0.9 for each class. The KNN classifier had the highest overall accuracy at 90.8% and was particularly strong at classifying VOID and DO events, achieving the highest F1 scores for DO and VOID events across all three classifiers. On the other hand, the SVM classifier scored the lowest overall CV accuracy (82.5%) and F1 score for each class but was the most sensitive classifier to ABD events (0.979). The ANN's CV accuracy (88.8%) was comparable to that of the KNN classifier. Furthermore, despite being less sensitive to VOID/DO events and ABD events compared to the KNN and SVM, respectively, the ANN achieved the best ROC AUC for all four classes, suggesting that this classifier generalized most effectively to all types of urological events.

These results suggested that the ANN classifier was the best candidate for real-time single-sensor urological event classification. Despite the strong classification ability of the KNN with the fewest number of features (17), the KNN may not be feasible to implement in low-power hardware as all training data must be stored on the device to make an inference on newly received data. While the storage overhead of the SVM classifier is quite possibly the lowest across all three models, its classification accuracy and ability to generalize to all types of urological events are the weakest. Therefore, the ANN model best balanced storage demands and performance. Although a higher number of features must be computed, the sole storage requirements for the host device are the neural network's pretrained weights and biases. Specifically, for real-time implementation, the neural network can be trained on a capable computing machine such as a PC. Then, a low-power device can read the weights and biases from onboard storage and execute matrix math computations to make an inference on new data. Ongoing work is focused on validating the performance of the ANN model on a Teensy 3.6 microcontroller for inference from real-time P_{VES} data, implemented using ARM matrix math libraries.

5 Conclusion

This chapter evaluated two potential future approaches for vesical pressure signal processing during filling cystometry. Specifically, these approaches are feasible for real-time application with minimal delay. This is critical, as clinical examination requires feedback between operator and patient to detect and fix common issues arising from movement, catheter mis-position, frequent detrusor activity caused by the catheters or filling rate, etc. While we did not demonstrate results from real-time implementation, both approaches employ causal signal processing chains with minimal delay. Further, both approaches focused on evaluating bladder activity only from a single measurement channel in the bladder (P_{VES}), rather than the traditional two-channel approach.

We demonstrated that single-catheter cystometry using a wavelet-based reconstruction estimator reproduced actual P_{DET} pressures from P_{VES} measurements. Waveform corruption introduced by abdominal pressure (P_{ABD}) was reduced first by least-mean-square fitting of reconstruction weights and, second, by a simple event detection scheme. The event detection was developed to identify the two most common artifacts present in P_{ABD} recordings (cough and Valsalva). Very slow, long-duration abdominal pressure events, however, were not effectively detected by this approach, which is a limitation imposed by the need for real-time operation without an artificial delay buffer.

Next, we demonstrated that supervised machine learning classifiers are effective tools for identifying bladder events, even when only a single channel (P_{VES}) measurement is used. Three supervised learning models were trained to operate on eight-sample segments in order to increase the data range for feature computation, while still demonstrating classification capability over short windows of data—which is critical for real-time implementation in the future. The evaluated classifiers (KNN, SVM, and ANN) were cross-validated on a relatively small dataset of 16,000 segments and showed accurate identification of specific bladder events (abdominal, bladder voiding, and detrusor overactivity). The supervised learning approach presented here may be feasible in a traditional urodynamics environment for automated event annotation. Furthermore, if implemented for real-time use, such techniques can improve the functionality of emerging catheter-free sensors and neuro-prosthetic systems for closed-loop bladder control. Future approaches may include investigating advanced deep learning models to remove the need for intermediate feature selection steps.

Acknowledgments SRS Medical has partially funded this research (Dr. Damaser, PI) and licensed relevant IP from the Cleveland Clinic. Drs. Damaser and Majerus are inventors on that IP and have significant financial interests in the outcome of this project. Mr. Brody is employed by SRS Medical and has significant financial interests in this project. This work does not represent the views of the US Department of Veterans Affairs or the US Government.

References

1. P. Abrams *et al.*, "The standardisation of terminology in lower urinary tract function: report from the standardisation sub-committee of the International Continence Society," *Urology*, vol. 61, no. 1, pp. 37–49, Jan. 2003, doi: https://doi.org/10.1016/s0090-4295(02)02243-4.
2. N. N. Maserejian *et al.*, "Incidence of lower urinary tract symptoms in a population-based study of men and women," *Urology*, vol. 82, no. 3, pp. 560–564, Sep. 2013, doi: https://doi.org/10.1016/j.urology.2013.05.009.
3. J. Nordling, "Functional assessment of the bladder," *Ciba Found. Symp.*, vol. 151, pp. 139–147; discussion 147–155, 1990, doi: https://doi.org/10.1002/9780470513941.ch8.
4. I. Clausen, L. G. W Tvedt, and T. Glott, "Measurement of Urinary Bladder Pressure: A Comparison of Methods," *Sensors*, vol. 18, no. 7, p. 2128, Jul. 2018, doi: https://doi.org/10.3390/s18072128.
5. S. Leslie, P. Tadi, and M. Tayyeb, "Neurogenic Bladder and Neurogenic Lower Urinary Tract Dysfunction," *StatPearls*, Nov. 2022, Accessed: Apr. 10, 2023. [Online]. Available: https://www.statpearls.com/articlelibrary/viewarticle/25789/
6. E. S. van Waalwijk van Doorn, A. H. Meier, A. W. Ambergen, and R. A. Janknegt, "Ambulatory urodynamics: extramural testing of the lower and upper urinary tract by Holter monitoring of cystometrogram, uroflowmetry, and renal pelvic pressures," *Urol. Clin. North Am.*, vol. 23, no. 3, pp. 345–371, Aug. 1996, doi: https://doi.org/10.1016/s0094-0143(05)70317-7.
7. E. van Waalwijk van Doorn *et al.*, "Standardisation of ambulatory urodynamic monitoring: Report of the Standardisation Sub-Committee of the International Continence Society for Ambulatory Urodynamic Studies," *Neurourol. Urodyn.*, vol. 19, no. 2, pp. 113–125, 2000, doi: https://doi.org/10.1002/(sici)1520-6777(2000)19:2<113::aid-nau2>3.0.co;2-#.
8. S. C. Radley, D. J. Rosario, C. R. Chapple, and A. G. Farkas, "Conventional and ambulatory urodynamic findings in women with symptoms suggestive of bladder overactivity," *J. Urol.*, vol. 166, no. 6, pp. 2253–2258, 2001, doi: https://doi.org/10.1016/S0022-5347(05)65545-0.
9. M. Serati, A. Braga, P. F. W. M. Rosier, S. de Wachter, A. Uren, and E. Finazzi-Agrò, "Acceptability and perceived value of urodynamics from the patient perspective: A narrative review," *Neurourol. Urodyn.*, vol. 41, no. 5, pp. 1065–1073, Jun. 2022, doi: https://doi.org/10.1002/nau.24932.
10. M. J. Drake, S. K. Doumouchtsis, H. Hashim, and A. Gammie, "Fundamentals of urodynamic practice, based on International Continence Society good urodynamic practices recommendations," *Neurourol. Urodyn.*, vol. 37, no. S6, pp. S50–S60, 2018, doi: https://doi.org/10.1002/nau.23773.
11. B. Abelson, S. Majerus, D. Sun, B. C. Gill, E. Versi, and M. S. Damaser, "Ambulatory urodynamic monitoring: state of the art and future directions," *Nat. Rev. Urol.*, vol. 16, no. 5, pp. 291–301, 2019, doi: https://doi.org/10.1038/s41585-019-0175-5.
12. J. Melgaard and N. J. M. Rijkhoff, "Detecting Urinary Bladder Contractions: Methods and Devices," *J. Sens. Technol.*, vol. 04, no. 04, pp. 165–176, 2014, doi: https://doi.org/10.4236/jst.2014.44016.

13. T. L. Frenkl *et al.*, "Variability of urodynamic parameters in patients with overactive bladder," *Neurourol. Urodyn.*, vol. 30, no. 8, pp. 1565–1569, Nov. 2011, doi: https://doi.org/10.1002/nau.21081.
14. K. T. Hobbs *et al.*, "Machine Learning for Urodynamic Detection of Detrusor Overactivity," *Urology*, vol. 159, pp. 247–254, Jan. 2022, doi: https://doi.org/10.1016/j.urology.2021.09.027.
15. S. J. A. Majerus *et al.*, "Feasibility of Real-Time Conditional Sacral Neuromodulation Using Wireless Bladder Pressure Sensor," *IEEE Trans. Neural Syst. Rehabil. Eng.*, vol. 29, pp. 2067–2075, 2021, doi: https://doi.org/10.1109/TNSRE.2021.3117518.
16. M. A. Soebadi, M. Bakula, L. Hakim, R. Puers, and D. de Ridder, "Wireless intravesical device for real-time bladder pressure measurement: Study of consecutive voiding in awake minipigs," *PLoS ONE*, vol. 14, no. 12, p. e0225821, 2019, doi: https://doi.org/10.1371/journal.pone.0225821.
17. M. A. Soebadi, T. Weydts, L. Brancato, L. Hakim, R. Puers, and D. De Ridder, "Novel implantable pressure and acceleration sensor for bladder monitoring," *Int. J. Urol.*, vol. 27, no. 6, 2020, doi: https://doi.org/10.1111/iju.14238.
18. I. Bentellis, S. Guérin, Z.-E. Khene, R. Khavari, and B. Peyronnet, "Artificial intelligence in functional urology: how it may shape the future," *Curr. Opin. Urol.*, vol. 31, no. 4, p. 385, Jul. 2021, doi: https://doi.org/10.1097/MOU.0000000000000888.
19. Z. E. Cullingsworth *et al.*, "Automated quantification of low amplitude rhythmic contractions (LARC) during real-world urodynamics identifies a potential detrusor overactivity subgroup," *PLOS ONE*, vol. 13, no. 8, p. e0201594, Aug. 2018, doi: https://doi.org/10.1371/journal.pone.0201594.
20. T. Niederhauser *et al.*, "Detection and quantification of overactive bladder activity in patients: Can we make it better and automatic?," *Neurourol. Urodyn.*, vol. 37, no. 2, pp. 823–831, 2018, doi: https://doi.org/10.1002/nau.23357.
21. R. Karam *et al.*, "Real-Time Classification of Bladder Events for Effective Diagnosis and Treatment of Urinary Incontinence," *IEEE Trans. Biomed. Eng.*, vol. 63, no. 4, pp. 721–729, Apr. 2016, doi: https://doi.org/10.1109/TBME.2015.2469604.
22. H.-H. S. Wang, D. Cahill, J. Panagides, C. P. Nelson, H.-T. Wu, and C. Estrada, "Pattern recognition algorithm to identify detrusor overactivity on urodynamics," *Neurourol. Urodyn.*, vol. 40, no. 1, pp. 428–434, 2021, doi: https://doi.org/10.1002/nau.24578.
23. R. Karam, S. J. A. Majerus, D. J. Bourbeau, M. S. Damaser, and S. Bhunia, "Tunable and Lightweight On-Chip Event Detection for Implantable Bladder Pressure Monitoring Devices," *IEEE Trans. Biomed. Circuits Syst.*, vol. 11, no. 6, pp. 1303–1312, 2017, doi: https://doi.org/10.1109/TBCAS.2017.2748981.
24. H. K. Awada, P. C. Fletter, P. J. Zaszczurynski, M. A. Cooper, and M. S. Damaser, "Conversion of urodynamic pressures measured simultaneously by air-charged and water-filled catheter systems," *Neurourol. Urodyn.*, vol. 34, no. 6, pp. 507–512, 2015, doi: https://doi.org/10.1002/nau.22633.
25. P. Gans and J. B. Gill, "Examination of the Convolution Method for Numerical Smoothing and Differentiation of Spectroscopic Data in Theory and in Practice," *Appl. Spectrosc.*, vol. 37, no. 6, pp. 515–520, Nov. 1983, doi: https://doi.org/10.1366/0003702834634712.
26. Abraham. Savitzky and M. J. E. Golay, "Smoothing and Differentiation of Data by Simplified Least Squares Procedures.," *Anal. Chem.*, vol. 36, no. 8, pp. 1627–1639, Jul. 1964, doi: https://doi.org/10.1021/ac60214a047.
27. M. Akay and C. Mello, "Wavelets for biomedical signal processing," in *Proceedings of the 19th Annual International Conference of the IEEE Engineering in Medicine and Biology Society. "Magnificent Milestones and Emerging Opportunities in Medical Engineering" (Cat. No.97CH36136)*, Oct. 1997, pp. 2688–2691 vol.6. doi: 10.1109/IEMBS.1997.756888.
28. A. al-Qerem, F. Kharbat, S. Nashwan, S. Ashraf, and khairi blaou, "General model for best feature extraction of EEG using discrete wavelet transform wavelet family and differential evolution," *Int. J. Distrib. Sens. Netw.*, vol. 16, no. 3, p. 155014772091 1009, Mar. 2020, doi: https://doi.org/10.1177/1550147720911009.

29. J. S. Sahambi, S. N. Tandon, and R. K. P. Bhatt, "Using wavelet transforms for ECG characterization. An on-line digital signal processing system," *IEEE Eng. Med. Biol. Mag.*, vol. 16, no. 1, pp. 77–83, Jan. 1997, doi: https://doi.org/10.1109/51.566158.
30. E. Park, J. W. Lee, M. Kang, K. Cho, B. H. Cho, and K. S. Lee, "Detecting bladder biomarkers for closed-loop neuromodulation: A technological review," *Int. Neurourol. J.*, vol. 22, no. 4, pp. 228–236, Dec. 2018, doi: https://doi.org/10.5213/inj.1836246.123.
31. A. J. Chashmi and M. C. Amirani, "An Efficient and Automatic ECG Arrhythmia Diagnosis System using DWT and HOS Features and Entropy- Based Feature Selection Procedure," *J. Electr. Bioimpedance*, vol. 10, no. 1, pp. 47–54, Aug. 2019, doi: https://doi.org/10.2478/joeb-2019-0007.
32. D. Cvetkovic, E. D. Übeyli, and I. Cosic, "Wavelet transform feature extraction from human PPG, ECG, and EEG signal responses to ELF PEMF exposures: A pilot study," *Digit. Signal Process.*, vol. 18, no. 5, pp. 861–874, Sep. 2008, doi: https://doi.org/10.1016/j.dsp.2007.05.009.
33. R. J. Urbanowicz, M. Meeker, W. La Cava, R. S. Olson, and J. H. Moore, "Relief-based feature selection: Introduction and review," *J. Biomed. Inform.*, vol. 85, pp. 189–203, Sep. 2018, doi: https://doi.org/10.1016/j.jbi.2018.07.014.
34. A. Bansal and A. Singhrova, "Performance Analysis of Supervised Machine Learning Algorithms for Diabetes and Breast Cancer Dataset," in *2021 International Conference on Artificial Intelligence and Smart Systems (ICAIS)*, Mar. 2021, pp. 137–143. doi: https://doi.org/10.1109/ICAIS50930.2021.9396043.
35. A. K. Nandi, "From Multiple Independent Metrics to Single Performance Measure Based on Objective Function," *IEEE Access*, vol. 11, pp. 3899–3913, 2023, doi: https://doi.org/10.1109/ACCESS.2023.3235014.
36. S. Akter *et al.*, "Comprehensive Performance Assessment of Deep Learning Models in Early Prediction and Risk Identification of Chronic Kidney Disease," *IEEE Access*, vol. 9, pp. 165184–165206, 2021, doi: https://doi.org/10.1109/ACCESS.2021.3129491.

Deep Recurrent Architectures for Neonatal Sepsis Detection from Vital Signs Data

Antoine Honoré, Henrik Siren, Ricardo Vinuesa, Saikat Chatterjee, and Eric Herlenius

1 Introduction to Neonatal Sepsis Detection

We introduce the clinical and technical background of neonatal sepsis detection. We start by discussing prematurity and sepsis in the intensive care units (NICUs). We then embark on discussing feature extraction from vital signs monitors readily available in most modern NICUs. Finally, we introduce how supervised learning methods can be used to perform neonatal sepsis detection.

A. Honoré (✉) · S. Chatterjee
Division of Information Science and Engineering, KTH Royal Institute of Technology, Stockholm, Sweden

Department of Women's and Children's Health, Karolinska Institutet, Stockholm, Sweden
e-mail: honore@kth.se; sach@kth.se

H. Siren
FLOW, Engineering Mechanics, KTH Royal Institute of Technology, Stockholm, Sweden

Department of Women's and Children's Health, Karolinska Institutet, Stockholm, Sweden
e-mail: hsiren@kth.se

R. Vinuesa
FLOW, Engineering Mechanics, KTH Royal Institute of Technology, Stockholm, Sweden
e-mail: rvinuesa@kth.se

E. Herlenius
Department of Women's and Children's Health, Karolinska Institutet, Stockholm, Sweden
e-mail: eric.herlenius@ki.se

A. Ahmed, J. Picone (eds.), *Machine Learning Applications in Medicine and Biology*,
https://doi.org/10.1007/978-3-031-51893-5_5

1.1 Clinical Background

Neonatal Sepsis in Preterms

In this section, we introduce some background on the population of premature infants in the NICU. We then embark on discussing the physiological challenges that this population of patients experience during maturation. Then, we focus on the background related to sepsis, a clinical adverse event that may worsen the maturation of infants.

Prematurity

Before birth, a baby is dependent upon functions from the mother. After birth, the baby's functions must transition to adapt to a new environment. The lungs must now breath in air, the immune system must work on its own, the digestive system must learn to process food and excrete waste by itself. In some cases, the baby has difficulties with this transition process. This can be due to several factors, for instance, after a preterm or a difficult term birth. The normal gestation period for a baby is about 40 weeks long. Live birth occurring prior to 37 weeks of pregnancy is considered preterm birth. In this section, we give some background about the population of patients under scrutiny in this dissertation: premature infants. More specifically, we describe prematurity clinically, with details concerning the potential causes, the degrees of severity, and an overview of the epidemiology of prematurity. Evaluation of the sepsis risk and the preterm infants maturity is crucial in the decision-making in neonatology [1, 2].

The incidence of preterm birth (PB) is estimated to be around 11% worldwide [3]. There are differences between lower-income and higher-income countries. According to the World Bank, lower-income countries are countries with a gross national income (GNI) per capita of less than 1026 USD [4]. This represents a total population of 520 million across 21 countries. Higher-income countries are countries with a GNI per capita of more than 12,476 USD. This represents a total population of 1.080 billion across 77 countries. It is estimated that of all live births, 12–15% are preterm in lower-income countries and 9% in higher-income countries [5]. In Europe, the rate is around 6.2% and is even lower, 5%, in northern European countries [6, 7]. High incidence rate of preterm can however be found in both higher- and lower-income countries. In Northern America and Africa, the preterm birth rate is 11.9% and 10.6% respectively [6]. Of all the 113,077 live births registered in Sweden in 2020, 6502 (5.7%) were preterm [8].

Cardiorespiratory Interactions in Maturing Infants

The cardiorespiratory system is composed of two subsystems: the cardiovascular and the pulmonary system. These two subsystems interact to help deliver oxygen and other materials to different parts of the body and to carry cellular waste out of the body. The cardiovascular system is responsible for the transport of gases, nutrients, and cellular wastes in appropriate parts of the body via a network of arteries, veins, and capillaries. The pulmonary system is responsible for exchanging gases, such as oxygen and carbon dioxyde, between inhaled air and blood.

Complex interactions between the main organs of the cardiorespiratory system, namely the heart and the lungs, are essentials for maintaining the body and its organs into a stable homeostatic state. One example of such interactions is respiratory sinus arrhythmia, which is the increase in heart frequency during inhalation and the decrease in heart frequency during exhalation [9]. Cardiorespiratory interactions are mediated by the autonomic nervous system (ANS), which in the case of premature birth is ill-prepared for the transition from intra-uterine to extra-uterine milieu and thus cannot mediate complex cardiovascular interactions [10]. The increase in observable cardiovascular interactions was shown to be correlated with maturing ANS, in turn being correlated with healthier infants, and better outcome [11]. The ANS however matures differently in an extrauterine milieu compared to the "natural" intrauterine milieu. Preterm birth was shown to lead to underdeveloped ANS a few weeks after birth in equivalent term age case controls [12]. This was also confirmed to persist up to several months after birth [13]. The degraded ANS maturation in part explains why prematurity remains a burden for the infants, after hospital discharge, and throughout life.

A number of adverse clinical conditions have been shown to further alter ANS maturation [14], including necrotizing enterocolitis [15], intraventricular hemorrhage [16], and sepsis [17]. These conditions are frequent in NICUs and further worsen the impairment in ANS maturation caused by prematurity.

Neonatal Sepsis

We focus on the alteration of ANS maturation induced by sepsis. In this section, we discuss neonatal sepsis in terms of its incidence, short-term and long-term consequences, and its diagnosis.

Characterization and Incidence

Neonatal sepsis refers to a pathogen invasion into the bloodstream of infants less than 28 days old. Neonatal sepsis is generally defined differently depending upon the transmission pathway, the pathogen, and the number of post-natal days after which it is presented [18]. Early-onset sepsis (EOS) refers to sepsis diagnosed within 72 hours of life. It is generally caused by the transmission of a pathogen from the genito-urinary system to the infant during delivery or to the fetus during pregnancy. The risk of early-onset sepsis can be increased by various maternal factors, such as chorioamnionitis, colonization with Group B Streptococcus (GBS), preterm delivery, and prolonged rupture of membranes exceeding 18 hours [19]. Late-onset sepsis (LOS), as opposed to the EOS, refers to the condition resulting from a pathogen transmitted from the environment to the infant after delivery. It was shown on a large German cohort that intubation interventions increase the risk of infections in VLBW infants [20]. Similarly, a study in Singapore revealed that prolonged respiratory support was associated with an increased risk of LOS in VLBW infants [21]. Some LOS might also be late manifestations of pathogens transmitted during delivery or in utero.

The incidence of sepsis is higher in preterm than it is in term neonates [22]. Premature infants requiring particular invasive procedure including intravascular catheter insertion and endotracheal tubes are at increased risk of pathogen transmission. Deficient immune system and in particular immature epithelial barrier also increase the risk for adverse consequences of seemingly benign infections [23]. The rate of EOS in newborns is lower than the rate of LOS, mainly due to improvement in systematic pathogen screening [24, 25]. Worldwide and in the last decade (2009–2018), the incidence of sepsis was estimated at around 3.9% of all live-birth [26].

Mortality, Severity, and Comorbidities

The mortality in neonates affected by sepsis was estimated at 3.9% [26]. The mortality was also reported higher in EOS and in LOS and varies depending upon the pathogen. For instance, it was shown that E. coli pathogen is associated with higher mortality rates than GBS [27]. The severity of sepsis in neonates is dependent upon a variety of factors, including genetic factors, sex, age at birth, and pathogen-specific characteristics [28, 29]. Infants born prematurely have a less mature cardiorespiratory system and have a lower acquired immunity, which puts them at risk of overwhelming clinical signs of sepsis, leading to shock or death [30]. The immune response to pathogens often involves acceleration of the heart rate as a response to complex signaling pathways. The decreased elasticity in the heart due to immaturity impairs the ability of the heart to maintain cardiac output and consequently tissue perfusion to an appropriate level to the organ by increasing stroke volume [31]. Immaturity of the heart can thus lead to adverse outcomes such as organ dysfunction, with severity correlated to the degree of immaturity of the heart. Similarly, the immaturity of the cardiorespiratory system puts patients suffering from sepsis at risk of chronic lung disease into adulthood [32, 33].

Clinical Praxis for Diagnosis

Neonates can be asymptomatic and have a normal physical examination while bacteria are present in the blood. Thus, the evaluation of blood sample–based biomarkers plays an important role in neonatal sepsis diagnosis. The classical characterization of sepsis requires a positive blood culture, obtained following the observation of clinical signs of infections. The diagnosis of sepsis is difficult in practice because of the small quantity of blood that can be drawn from premature infants. Other definitions of sepsis have been proposed, relying on specific threshold-based criterions for heart rate, respiratory rate, blood pressure, and temperature [34]. The consensus for these criteria however excluded preterm infants, due to the highly variable vital signs patterns they exhibit during maturation [35]. For adults, the newest guidelines recommend the use of sequential organ failure assessment (SOFA) score [36]. In [37], the authors propose a deep learning–based SOFA score and show that they obtain better results than previously reported for 6 hours ahead of prediction of sepsis in adults. An extension of the adult SOFA score to neonates (nSOFA) was developed [38]. This score was initially shown to be predictive for mortality in very low birth weight patients infected with LOS in multiple studies [38]. nSOFA was then also shown useful to evaluate the risk of various adverse clinical conditions [39]. The score is thus sometimes used in practice for general patient assessment, but the standard for sepsis diagnosis in preterm infants remains a combination of blood sample–based biomarkers, including C-reactive protein (CRP), cerebrospinal fluid (CSF), and culture [40].

Preterm newborns admitted to NICUs undergo frequent blood sampling for diagnosis, monitoring, and management of illnesses. In critically sick neonates, the total volume of extracted blood may be as much as 58% of the total blood volume during the first weeks of life [41–43]. Blood loss due to sampling is considered a major cause of anemia in preterm infants. In extremely preterm newborns, anemia induced by excessive blood loss is suspected to be associated with the development of illnesses such as necrotizing enterocolitis and bronchopulmonary dysplasia [44]. Limiting the amount of blood drawn from sick neonates is therefore crucial to limit the risk for complications. However, too small volumes often lead to false-negative blood cultures, which restricts the precision and reliability of blood cultures for sepsis diagnosis. Other blood sampling–based markers including CRP can be used to detect the onset of an infection [45]. Urine samples may also be considered for assessing the presence of sepsis [46]. However, although laboratory evaluation was shown to have some predictive power, these methods are usually slow to rise [47] and by definition do not permit to identify the causative organism. CSF-based laboratory analysis can help reveal both the blood stream infection onset and the causative organism. CSF samples are obtained via a procedure named lumbar puncture, also known as spinal tap. This procedure consists of sampling fluid directly from spinal cord, making CSF samples more accurate markers to reveal ongoing involvement of the ANS in a systemic infection. CSF-derived molecule levels are however difficult to interpret in preterm neonates, with reference values varying for each patient.

1.2 Technical Background

Vital Signs Monitoring

This section focuses on the sensors required to perform vital signs monitoring. We discuss the monitoring of respiration, oxygen saturation, and heart rate, using monitors readily available in most modern NICUs.

Cardiorespiratory Interactions Impairment

Changes in vital signs, such as body temperature, respiratory drive, heart rate, and blood pressure, are a common response to sepsis in newborns. These changes occur following the releases of various inflammatory agents by the immune system, in response to pathogens entering the body. These inflammatory agents can alter the ANS, and this can manifest as changes in vital signs. The autonomic nervous system plays a crucial role in the connection between inflammation and changes in vital signs during sepsis. When pathogens enter the body, immune cells, cytokines, and microbial toxins quickly activate the ANS. The sympathetic nervous system responds by releasing catecholamines and stimulating the adrenal gland in an effort to eliminate the pathogen, which can cause symptoms such as a fast heart rate, rapid breathing, and changes in blood pressure. These changes may be subtle at first, but can become more severe as the sepsis progresses. Monitoring vital signs continuously can help detect changes in the newborn's condition before it becomes critical and provide additional information when clinical signs are not clear [17, 48].

Abnormalities are visible on the signals monitoring the heart rate, the respiratory frequency, and the level of oxygen saturation in the blood. In this section, we review these three vital signs and provide details about the underlying sensors and their pitfalls in clinical practice.

Arterial Oxygen Saturation Level from Pulse Oximetry

The gold standard to measure arterial oxygen saturation level, SaO_2, is the use of co-oxymetry performed on a blood sample [49]. Pulse oxymetry is a method to noninvasively estimate the level of peripheral saturation in oxygen (SaO_2) in the blood. Pulse oximeters measure oxygen saturation indirectly, via the analysis of photoplethysmograms (PPG) [50, 51]. PPGs are plethysmograms obtained optically by measuring the changes in light absorption of a tissue, see Fig. 1 for an example depiction. The volume of blood in perfused tissues varies drastically between heart beats. This in turn modifies the light absorption properties in the tissue and makes PPGs a good tool to retrieve information about the heart frequency, and other circulatory conditions, including the respiratory frequency and the SpO_2 [52].

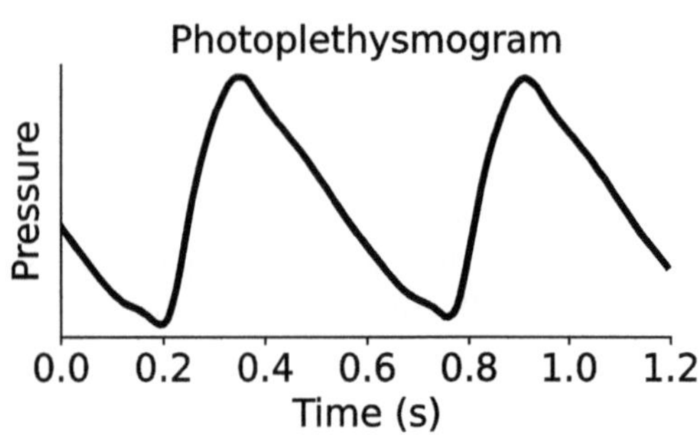

Fig. 1 Example PPG

Let us assume that arterial blood is only composed of deoxygenated hemoglobine (Hb) and oxygenated hemoglobine (HbO_2). In this context, Hb and HbO_2 are the only two sources of light absorption in arterial blood. The technique then relies on the fact that [Hb] and [HbO_2] have different absorption spectra in the red and near-infrared regions. By denoting [Hb] (resp. [HbO_2]) the concentration in deoxygenated (resp. oxygenated) hemoglobine, the saturation in oxygen (SaO_2) is then defined as:

$$\mathrm{SaO_2} = \frac{[\mathrm{HbO_2}]}{[\mathrm{HbO_2}] + [\mathrm{Hb}]} \text{ (in \%)}. \tag{1}$$

By denoting ϵ_0 (resp. ϵ_D) the absorption coefficient for deoxygenated (resp. oxygenated) hemoglobin, the total absorption coefficient ϵ is related to the SaO_2 as follows:

$$\epsilon = \epsilon_0 \mathrm{SaO_2} + \epsilon_D (1 - \mathrm{SaO_2}). \tag{2}$$

Thus, knowing ϵ_0 and ϵ_D, measuring ϵ allows assessing SaO_2 as follows:

$$\mathrm{SaO_2} = \frac{\epsilon - \epsilon_D}{\epsilon_0 - \epsilon_D} \tag{3}$$

In reality, tissues have other sources of light absorption in red and infrared regions, the most prominent one being the venous blood. To account for that, pulse oxymetrie was developed to specifically measure the increase in light absorption due to the increase in arterial blood volume following systole (heart compression). For this, commercial pulse oximetry measures PPGs in two different wavelengths, usually in the red and infrared. The use of two wavelengths is required to account for any interference caused by other pigments in the skin or to filter out any noise or interference from other sources, such as ambient light [53]. Then, the amplitude (AC) and the offset (DC) are extracted from the PPGs over a cardiac cycle and used to compute a ratio of ratios, R, defined as:

$$R = \frac{(AC/DC)_1}{(AC/DC)_2}. \tag{4}$$

However, the difference in refractive indexes between plasma and red blood cells significantly increases the probability of absorption of light in the blood and thus, SaO_2 cannot be estimated directly by measuring R. Instead, commercial pulse oximeter uses measurements of R and blood gas analysis based measurement of SaO_2 obtained from healthy controls, and estimates SaO_2 via a regression model:

$$\mathrm{SaO_2} = \frac{k_1 - k_2 R}{k_3 - k_4 R}, \tag{5}$$

where k_1, k_2, k_3, k_4 are scalar parameters. This process of finding the parameters by best-fit analysis is called the calibration process. The estimate of SaO_2 obtained via pulse oximetry is denoted SpO_2. The variability between control subjects used for calibration and the subjects under monitoring in clinical practice causes discrepancy between the true SaO_2 and the measured SpO_2. In newborn, the detection of light through tissues limits the areas where the light receiver and transmitter can be placed to fingertips and feets. In these areas, the sensor for PPGs is subject to movement artifact, which considerably degrades the estimation of SaO_2 [54]. Additionally, cases of poor peripheral perfusion conditions, e.g., low cardiac output or low surrounding temperature, lead to degraded estimations of SaO_2 [53].

Heart rate and respiratory rate could be extracted from PPG using frequency content analysis. These are however estimated from ECGs in neonates, due to the general unreliability of the PPGs in neonates.

Heart Rate from Electrocardiogram

Electrocardiograms (ECGs) are voltage recordings of the electrical activity of the heart [55]. ECGs are usually obtained from 12 electrodes (leads) placed on the skin. The patterns visible on ECGs allow tracking the changes in polarity of the heart during a cardiac cycle, see Fig. 2 for an example. The activity of the heart is generated by the periodic contraction and relaxation of heart compartments. The contraction and relaxation activity causes an ionic current flow and an electrical potential difference that is detectable using suitably placed electrodes. The signal resulting from the electrodes is called the electrocardiogram (ECG) signal. The heart contraction and relaxation pattern may be decomposed into three sub-pattern denoted as P, QRS, and T. These sub-patterns are mainly characterized by their shape, duration, and period. The period corresponds to the time between two successive occurences of the sub-pattern. The peak in electrical activity of the heart is known as the R-peak in the QRS pattern. The period of the R-peak is typically used to measure the inter-beat interval, also known as the RR-interval, and is used to give an instantaneous measure of the heart rate. The variation in the RR-interval (RRi) overtime, called heart rate variability (HRV), seems to be the most relevant information to investigate cardiac arrythmia and to determine the autonomic nervous

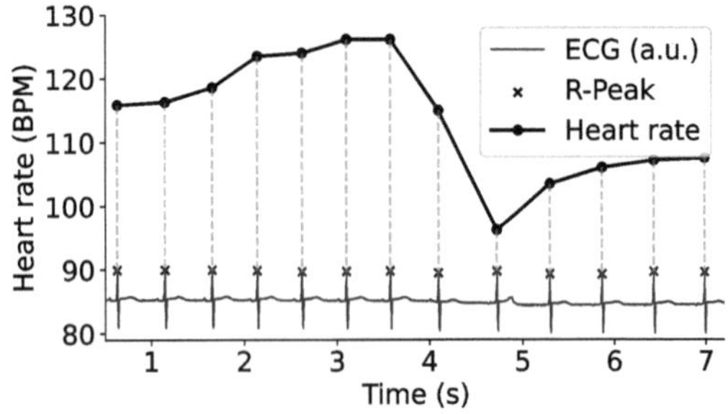

Fig. 2 Electrocardiogram with corresponding heart rate as derived from the R-peak period

system effect on the heart. The R-peak must be measured accurately when the ECG signal is to be used for patient assessment and disease diagnosis.

When the baby is young, the skin is irritable. For this reason, the staff tries to place as few ECG leads as possible. Also, the baby might be placed under a blanket for temperature control, and the blanket may scratch the leads off, causing disturbances in the recordings.

Respiration Rate from Chest Impedance

Respiration rate (RR) can be derived from chest impedance (CI) by detecting changes in electrical impedance caused by chest movements during spontaneous breathing [56] (Fig. 3). Sensors placed on the chest can measure changes in electrical impedance caused by the movement of the rib cage and diaphragm during inhalation and exhalation. The resulting impedance signal can then be used to calculate respiration rate. This technique is commonly used in medical and wearable devices to monitor respiration patterns. CI provides an indirect measure of spontaneous breathing, since it measures the changes in thoracic impedance due to lung aeration variations. RR measurements derived from CI can become inaccurate since impedance is also influenced by non-breathing-related chest wall movements [57]. Moreover, electrodes encompassing a small area on the chest were shown to produce RR measurements that are very sensitive to movement artifacts [58]. Increasing the number of electrodes and increasing the electrode area were shown to lead to more reliable RR measurements in the presence of motion artifacts [58]. However, these solutions have a limited impact for the preterm neonates, who have a small and shallow chest wall, leading to weak signals and low signal-to-noise ratio. Improvement of CI-based RR measurements was suggested at the algorithmic level instead [59, 60]. In particular, some works are focused on estimating the quality of the CI signal, along with the estimation of RR. This can be done using a combination of a physiological plausibility score of a period between two peaks of the CI and the correlation coefficient between an observed breathing pattern and the average breath morphology pattern [59]. Another approach consists of labeling true breathing patterns and to learn to discriminate true breathing patterns from artifacts patterns with classification algorithms [60].

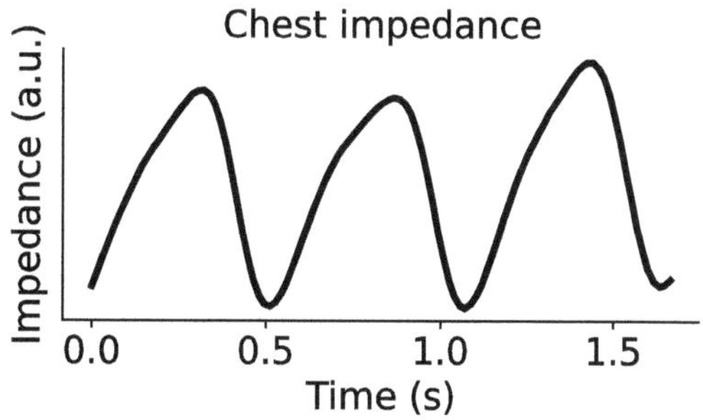

Fig. 3 Chest impedance waveform

Feature Extraction

In this section, we present the features we used to characterize vital sign signals. We divide the presentation into three parts: we first describe classical statistical moment-based features, then discuss the sample asymmetry, and lastly, the sample entropy, which is central to the characterization of cardiorespiratory coupling in preterms.

Current practice for sepsis detection from vital signs consists of measuring deterioration in cardiorespiratory coupling via features calculated on the vital signs data [11]. The features that have been investigated can generally be classified into two types: time domain and Fourier domain. Fourier domain features are typically used on the RR interval time series, to characterize the power in certain frequency bands [61]. Although the power in certain bands of the Fourier spectrum was shown to be associated with ANS maturation for the RR interval signal, no such associations have been shown for the respiratory rate and the blood pressure signal. Also, it is often possible to capture frequency domain information from time domain analysis of signals, e.g., recovering the spectral envelop of signals via auto-regressive process fitting. Moreover, a time domain feature named sample entropy, used as key components of the commercial HeRO monitor [62], was shown highly predictive of neonatal sepsis and was also shown to be associated with other physiological traits. Thus, we focused on the calculation and use of time domain features calculated on all the readily available vital signs, namely the inter-beat-interval (IBI), the oxygen saturation level in the blood (SpO_2), and the respiratory frequency (RF).

Statistical Moments

In statistics, moments are used to characterize the shape of a probability density function (pdf) of a random variable X (assume 1d for now) around it's mean. For a pdf f_X, the moment of order 1 corresponds to the expected value and is defined as:

$$m_1 = \mu = \mathbb{E}[X] = \int_{-\infty}^{+\infty} x f_X(x)dx, \tag{6}$$

and the moments of order $k \geq 2$ are defined:

$$m_k = \mathbb{E}\left[(X - \mathbb{E}[X])^k\right] = \int_{-\infty}^{+\infty} (x - \mathbb{E}[X])^k f_X(x)dx. \tag{7}$$

A bounded density function can be exactly determined by the infinite collection of its moments: $\{m_i\}_{i=1}^{\infty}$, given some conditions on the sequence of moments. In some cases, the first few moments are enough to exactly determine the pdf of a random variable. For instance, a Gaussian pdf is exactly determined by its mean and its variance, the moment of order 2 noted σ^2. The variance of a random variable is interpreted as the spread of its pdf around its mean. The moment of order 3, denoted

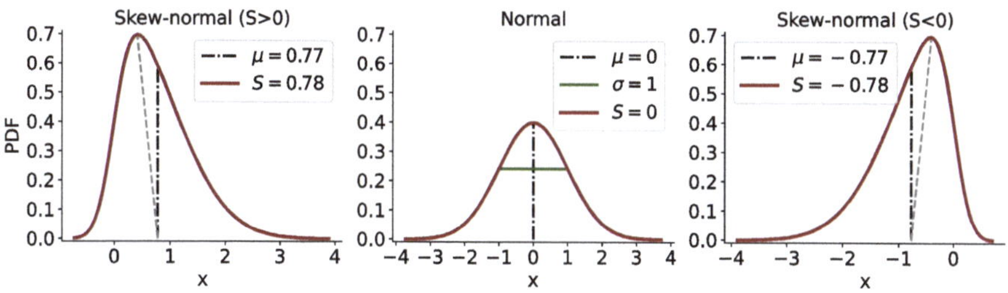

Fig. 4 Example PDF of skew-normal distributions with different values of skewness

as S, characterizes the skewness of a pdf around its expected value. Examples of pdf with various values of S are depicted in Fig. 4. The moment of order 4, denoted as ξ and called the Kurtosis, as opposed to the moment of lower order, does not characterize the shape of a pdf around its expected value, but rather characterizes the tail of a pdf, i.e., the behavior away from the expected value. This interpretation arises from (7), by noting that numbers <1 raised to the power $k = 4$ become very small and thus, do not contribute to the kurtosis, whereas numbers that are further away from the mean will become rapidly very large when raised to the power $k = 4$ and thus, will contribute more to the kurtosis. Thus, another interpretation of kurtosis is that it measures the propensity of a pdf to generate outliers.

In practice, the true f_X is unknown, and only a finite number N of realizations of the random variable X, $\{x_n\}_{n=1}^{N}$, are available. The realizations x_n are also called random *draws*, or random *samples* from the pdf f_X. Since f_X is unknown, the moments cannot be calculated directly from (6) and (7). Instead, they must be estimated using the samples drawn from the distribution. Empirical estimators of the moments are denoted with a $\hat{}$ and are called the sample moments. The sample mean is calculated as:

$$\hat{m}_1 = \hat{\mu} = \frac{1}{N}\sum_{i=1}^{N} x_i, \tag{8}$$

and the sample moments of order $k \geq 2$:

$$\hat{m}_k = \frac{1}{N}\sum_{i=1}^{N} \left(x_n - \hat{\mu}\right)^k. \tag{9}$$

To allow the comparison of the coefficient across types of pdfs, in practice, the standardized moments are calculated, for the skewness this gives:

$$\hat{S} = \frac{\hat{m}_3}{\hat{m}_2^{3/2}}, \tag{10}$$

and for kurtosis:

$$\hat{\xi} = \frac{\hat{m}_4}{\hat{m}_2^2}. \tag{11}$$

The sample moments were used in neonatal sepsis detection studies, e.g., [63]. A complete review of the algorithms, features, and outcomes reported in neonatal sepsis detection studies was performed [64].

Sample Entropy

Sample entropy (SampEn) is a feature that was introduced in order to capture abnormal heart rate characteristics, which are known to occur early in the course of neonatal sepsis [65]. Sample entropy can be interpreted as the rate of generation of new information in a sample of length N. A reduced SampEn indicates an increased regularity and in turn reduced cardiorespiratory interactions often predictive of neonatal sepsis. SampEn aims at improving the previously established approximate entropy (ApEn) [66]. These methods rely on pattern matching to measure the regularity of a time series. For this, one must specify two parameters: $(m, r) \in \mathbb{N} \times \mathbb{R}$, used to define matching patterns of length m with tolerance r, i.e., a sequence of length m of consecutive samples where all the samples are within r from each other. SampEn is expressed as follows for a sequence of length N:

$$\text{SampEn}(N, m, r) = -\ln \frac{P_m}{P_{m+1}}, \tag{12}$$

where P_m denote the number of matching patterns of length m in the sequence.

In practice, the parameter r is often chosen as a fraction of the standard deviation of the data sequence [61, 65]. Most studies have used $r = 0.2\sigma$, where σ denote the standard deviation of the samples in the data sequence. The parameter m is chosen depending on dynamic properties of the data. Lake et al. propose to choose a value corresponding to the order of an AR process fitting the data accurately [65]. In the case of neonatal IBI signal, this corresponds to $m \geq 3$, and most studies in the literature have used the smallest, $m = 3$ [61].

Sample Asymmetry

Sample asymmetry (SampAsy) was introduced to characterize the asymmetry of the distribution of a subsequence of IBI data from neonates [67]. It was introduced to tackle the shortcomings, skewness, which also characterizes the asymmetry of a distribution around its mean. The skewness reports on the asymmetry of a distribution around the expected value, but the expected value is not a good reference point in asymmetric distributions. As is depicted on Fig. 4, the mean is off the mode of the distribution for asymmetric distributions. Sample asymmetry enables to measure asymmetry around any reference point, in particular around the median, which is closer to the mode of the skewed distributions. Also, median is less sensitive to outliers in the data, which is important for the study of IBI data. The sample asymmetry R is calculated as follows for N data samples $\{x_i\}_{i=1}^N$:

$$R = \frac{\sum_{x_i > \alpha} (x_i - \alpha)^2}{\sum_{x_i \leq \alpha} (x_i - \alpha)^2}, \tag{13}$$

where $\alpha = \text{median}(\{x_i\})$. When the data are distributed symmetrical around the median, we have $R = 1$. Values of R greater than 1 indicate more mass in values to the right of the median, while values of R less than 1 indicate more mass in values to the left of the median. In neonates, $R > 1$ was shown associated with more frequent accelerations of the heart rate and $R < 1$ with more frequent deceleration. Thus, the sample asymmetry is of particular interest when calculated on the IBI signal [17, 67].

Supervised Learning for Binary Classification

In this section, we describe the supervised learning methods we employed to perform neonatal detection in a binary classification framework. We first introduce the principle of predictive modeling, then describe how to practically approach the evaluation of the generalization abilities of a predictor. Finally we discuss how to deal with class imbalance, inherent to neonatal sepsis detection datasets.

Predictive Modeling

Predictive modeling refers to the use of statistical tools to predict outcomes from observations. Classification refers to the action of inferring a quantity, $y \in \mathcal{Y}$, with $\mathcal{Y}$ a finite set of labels, corresponding to a collection of d observations $\mathbf{x} = [x_1, \ldots, x_d]^T$ and taking values in a vector space $\mathcal{X}$. This is classically done by designing a deterministic function

$$f : \mathcal{X} \rightarrow \mathcal{Y}, \tag{14}$$

taking a set of observations as inputs, and computing the corresponding output $y = f(\mathbf{x})$. No randomness occurs when computing y from $\mathbf{x}$, i.e., the same output will always be computed given the same input. The quantity $\mathbf{x}$ is also called an input vector or an input sample. The quantity y is also called an output label. The function f is often unknown in practice and must be chosen so that it can computes correct labels given a set of input vectors. Suppose that we are given a set of N observations:

$$\mathcal{D} = \{(\mathbf{x}_i, y_i) \in \mathcal{X} \times \mathcal{Y}\}_{i=1}^N \mid i = 1, \ldots, N\}. \tag{15}$$

We are interested in finding a function f that maps each vector $\mathbf{x}_i$ to its corresponding label y_i.

Let us denote

$$l : \mathcal{Y}^2 \to \mathbb{R}^+, \tag{16}$$

a loss function, informally indicating whether two labels are close to or far from one another.

For instance, let us consider the binary loss function:

$$l(y, y') = \begin{cases} 0 & \text{if } y = y', \\ 1 & \text{otherwise.} \end{cases} \tag{17}$$

For a classification problem, we are interested in finding function f such that the sum of the loss encountered on all the elements in $\mathcal{D}$ is minimized. Finding f assuming that pairs of input and labels are available is referred to as performing supervised learning of function f. Often, f is specified together with a set of learnable parameters θ and denoted f_θ. The parameters can be adjusted to modify the way function f_θ transforms the input to compute the output y.

The internal transformations involved in the computation of y are what defines the architecture of function f_θ. The goal then becomes to find a set of parameters $\theta^\star$, which minimizes the loss encountered on $\mathcal{D}$, i.e., finding

$$\theta^\star = \arg\min_\theta \sum_i l(f_\theta(\mathbf{x}_i), y_i) \tag{18}$$

Solving the minimization problem (17) requires finding patterns in the input data samples enabling the computation of accurate labels. In other words, an algorithm taking as input a loss function l, a dataset $\mathcal{D}$, and an architecture f_θ, exploits correlations between the data and the labels in order to provide the best parameter set $\theta^\star$. In general, it is difficult to control the patterns found in the data by a learning algorithm. The correlations found can be very specific to the dataset provided for learning. In this case, the learnt function $f_{\theta^\star}$ will fail to predict accurate labels on new and unseen data samples $\mathbf{x}'$. This over-fitting problem prevents $f_{\theta^\star}$ from making relevant predictions in practice. Such a function is said to have bad *generalization* properties. To formally describe this, we revisit the empirical risk minimization with probability distributions.

Cross-Validation

Cross-validation is statistical framework developed to perform out-of-sample testing of a prediction model [68]. It was developed in order to evaluate empirically the generalization ability of a prediction model on samples it has not trained on. In general, cross-validation relies on a sub-sampling procedure of the available dataset

$\mathcal{D}$. The key idea is to split dataset $\mathcal{D}$ multiple times into a training set and a validation set. Out-of-sample testing results are then obtained by averaging the performances across all validation sets.

Binary Classification Performance Metrics

Binary classification is a prediction problem where a discrete label $y \in \{0, 1\}$ is to be predicted from observations $\mathbf{x}$. The performances of a candidate model for this prediction task can be evaluated using various performance metrics, measuring different aspects of the performances of the model. For a classification task, the prediction $\hat{y}$ of a model must be made discrete to allow comparison with the discrete label y. We summarize the different outcomes depending on the possible combinations of labels and predictions in Table 1.

These elementary scores report on the different aspects of the performances of a binary classification model. A classification model performs as good as the TN and TP scores are high, and the FP and FN scores are low. In order to compare different algorithms and different datasets, it is often useful to aggregate these scores. For instance, if one dataset has many more samples than another, the same TP count (say TP = 10) will not have the same meaning. To account for this, we define other metrics, summarized in Table 2.

All the metrics described above assume that the outputs of the prediction model are binary. In practice, however, it is often desirable to design models that output a label probability (i.e. $y \in [0, 1]$) rather than a discrete label value (i.e. $y \in \{0, 1\}$). Often, the discretization of the output of prediction models is done with a threshold,

Table 1 Elementary binary classification performance metrics

y	$\hat{y}$	Metric name	Conclusion
0	0	True negative (TN)	Correct
0	1	False positive (FP)	Type I error
1	0	False negative (FN)	Type II error
1	1	True positive (TP)	Correct

Table 2 Binary classification performance metrics

Name	Formula
Sensitivity	$\frac{TP}{TP+FN}$
Specificity	$\frac{TN}{TN+FP}$
Precision	$\frac{TP}{TP+FP}$
Accuracy	$\frac{TP+TN}{TP+TN+FP+FN}$
Balanced accuracy	$\frac{\text{Sensitivity}+\text{Specificity}}{2}$

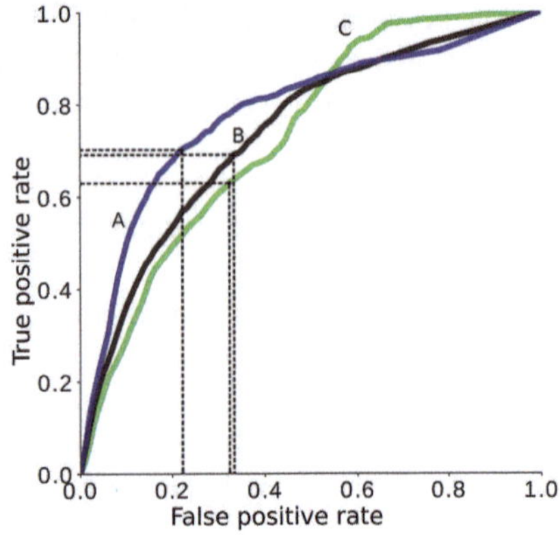

Fig. 5 ROC curves for three hypothetical models: A, B, C. [71, Fig 7 (a)]. The dashed vertical (resp. horizontal) lines correspond to the false-positive rate (resp. true positive rate) associated with the choice of an optimal threshold

denoted, for instance, as t, on the output label probability. This makes the computed scores threshold dependent, leading to misleading results when the models are poorly calibrated [69]. Some performance metrics such as the receiver operating characteristics (ROCs) have been designed to measure performances regardless of the choice of the threshold [70, 71]. ROC curves consist of plotting the true-positive rate (Sensitivity) versus the false-positive rate (1-Specificity) for different values of the threshold $t \in [0, 1]$. ROC is a threshold-independent visual representation of the performances of binary classification model [71]. This is exemplified in Fig. 5. The area under the receiver operating characteristic (AUROC) can be calculated and provides a threshold-independent quantity for performance assessment [71]. The true AUROC involves the integration of the ROC curve for all possible threshold values. In practice, the ROC is approximated with a finite number n of thresholds: $t_1, \ldots, t_n$. The calculation of AUROC is thus typically performed with the trapezoidal rule numerical integration method:

$$\text{AUROC} \approx \sum_i \frac{(\text{FPR}_{i+1} - \text{FPR}_i)(\text{TPR}_i + \text{TPR}_{i+1})}{2}, \tag{19}$$

where FPR_i and TPR_i are respectively the values of FPR and TPR obtained when the prediction output is thresholds with threshold t_i.

Cost-Sensitive Learning for Class Imbalance

Cost-sensitive learning is a framework to implement external sample-based constraints in a learning problem [72]. Class imbalance is a characteristic of supervised learning datasets. When a dataset is imbalanced, the number of samples differs across the classes. In such cases, performance metrics must be chosen adequately in order to truly assess the performances of predictive models (See Sect. 1.2). When predictive models are learnt discriminatively on imbalanced datasets, the loss function can be adapted in order to avoid learning a predictor biased toward predicting the label of the majority class. Deep architectures are usually trained with BP-based GD since the training objective is a scalar and is differentiable wrt to the trainable weights. In a supervised learning setting, the scalar objective is typically a sum of scalars, each corresponding to the objective function applied to individual samples. This means that penalty terms can be used in order to over-compensate for the false predictions made on the minority class. For imbalanced binary classification problems, the training objective is modified as follows:

$$L(\theta) = \sum_{i \in \{0,1\}} w_i L_i(\theta), \tag{20}$$

where for class i, $L_i(\theta) = \sum_{j \in \{j | y_j = i\}} l(f_\theta(\mathbf{x}_j), y_j)$ corresponds to the loss on the samples of class i, and w_i is the penalty weight. The penalty weights are chosen so that the weighted sum of the samples in one class is equal to the weighted sum of the samples in the other class. A natural choice for w_i is to use a quantity that is inversely proportional to the class frequency, typically:

$$w_i = \frac{N}{2\#_i}, \tag{21}$$

where N is the total number of samples and $\#_i$ denotes the number of samples of class i. The classical loss can be recovered by setting all $w_i = 1$.

Recurrent Models for Sequential Data

We describe the deep recurrent architectures we employed in our study. We first introduce the vanilla architecture and then discuss the improvements brought by the more advanced long-short-term memory architecture.

Introducing the Vanilla Recurrent Neural Network

RNNs are a family of deep architectures suited to tasks containing sequential data. These models are, in theory, able to capture nonlinearity and long-term dependencies in input time sequences. The model architecture can be formulated in two steps:

$$\mathbf{h}_t = \mathbf{g}_\theta(\mathbf{x}_t, \mathbf{h}_{t-1}), \quad \hat{y}_t = f_\phi(\mathbf{h}_t), \tag{22}$$

where θ denotes the set of trainable parameter for function $\mathbf{g}_\theta$, which returns the hidden state at time t, $\mathbf{h}_t$, from (1) the data at time t, $\mathbf{x}_t$, and (2) the hidden state at the previous time step, $\mathbf{h}_{t-1}$. In control theory, this model is equivalent to a state space model where $\mathbf{x}$ is the control variable. We depict the principle of an RNN in Fig. 6.

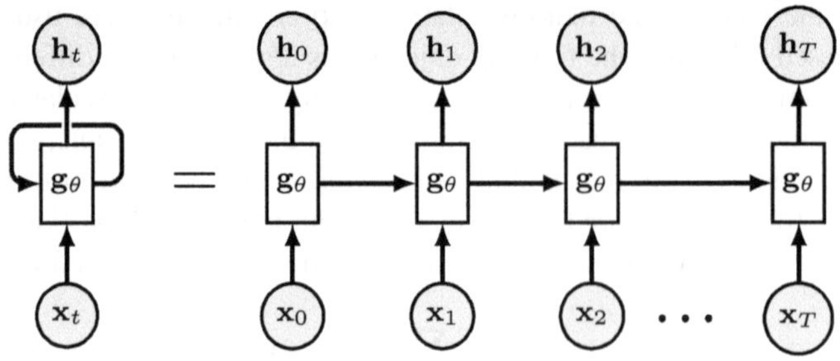

Fig. 6 Recursive process of RNN

ϕ denotes the set of trainable parameters for the measurement function f_ϕ. f_ϕ is a trainable function of the hidden state at time t returning the estimated probability of sepsis at time t, $\hat{y}_t$. The recurrent process modeled by $\mathbf{g}_\theta$ allows RNN models to capture information from previous time samples of observed data into the hidden state variable $\mathbf{h}_t$. The variable $\mathbf{h}_t$ is then used by the measurement model f_ϕ to predict an estimate of the desired observed outcome $\hat{y}_t$. The measurement model does not contain a recurrence, and its role is to map the hidden variable at time t to the predicted outcome variable at time t. It is often chosen as a fully connected architecture. In our present work, we chose a simple network with one hidden layer.

$$\hat{y}_t = f_\phi(\mathbf{h}_t) = \text{sigmoid}(\mathbf{w}_o^T \mathbf{h}_t + \mathbf{b}_o), \tag{23}$$

where $\phi = \{\mathbf{w}_o \in \mathbb{R}^h, \ \mathbf{b}_o \in \mathbb{R}\}$ denotes the set of the linear transform and the bias term for the sepsis probability computation. With such a simple measurement model, the process model should be chosen so as to be able to capture the complex dependencies in the input data, as well as being able to be trained for long sequences. The first architecture proposed for $\mathbf{g}_\theta$ was inspired from the work from [73]. The design is simply two additive linear layers and a nonlinear activation, as follows:

$$\mathbf{h}_t = \mathbf{g}_\theta(\mathbf{x}_t, \mathbf{h}_{t-1}) = \tanh(W_i \mathbf{x}_t + \mathbf{b}_i + W_s \mathbf{h}_{t-1} + \mathbf{b}_s), \tag{24}$$

where $\theta = \{\mathbf{b}_s \in \mathbb{R}^h, \ W_s \in \mathbb{R}^{h \times h}, \ W_i \in \mathbb{R}^{h \times d}, \ \mathbf{b}_i \in \mathbb{R}^h\}$ denotes the set of trainable bias vectors and weight matrices for the latent space vector computation. This RNN architecture has the advantage of being easily implemented and computationally efficient since the forward pass only consists of two linear transforms and a static nonlinearity. However, this architecture suffers from the vanishing or exploding gradient problem when it is trained with back-propagation. This has been discussed in detail in the literature, e.g., [74]. The problem arises from the recurrent nature of the architecture. Let us denote $w \in \mathbb{R}$ one of the components of matrices W_s or W_i. Informally, the gradient wrt w of a loss computed on a sequence of length T of predicted outcomes will depend upon the term w^T, the value of w raised to the power T, will either explode if $w > 1$, or vanish if $w < 1$,

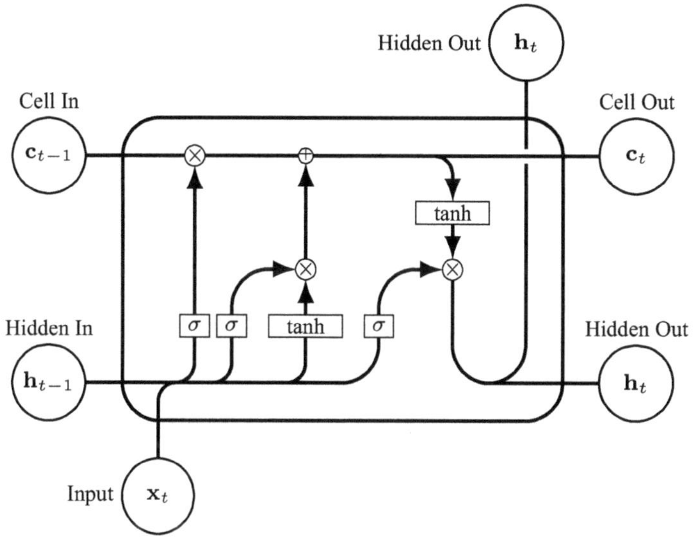

Fig. 7 Process function of a LSTM unit

when long sequences are processed, i.e., when the value of T becomes large. This was shown to be an issue for values of $T \geq 100$ approximately.

LSTM to Address Exploding and Vanishing Gradient

The problem of exploding/vanishing gradient arises from the architecture of the vanilla RNN. More advanced RNN architectures have been proposed to prevent the gradient of the loss to be poised by terms raised to the power T. One such architecture is the long-short-term memory (LSTM) unit [75]. The process function of an LSTM unit is depicted in Fig. 7.

LSTMs are designed with gates controlling the information flow in the process transform. The core feature of the LSTM compared to vanilla RNN is to embark a memory cell variable $\mathbf{c}_t$ in addition to the hidden state variable $\mathbf{h}_t$. The equations describing the process function of LSTM depicted in Fig. 7 are as follows:

$$\tilde{\mathbf{c}}_t = \tanh\left(\mathbf{W}_{cx}\mathbf{x}_t + \mathbf{W}_{ch}\mathbf{h}_{t-1} + \mathbf{b}_c\right), \tag{25}$$

$$\Gamma_f = \sigma\left(\mathbf{W}_{xf}\mathbf{x}_t + \mathbf{W}_{hf}\mathbf{h}_{t-1} + \mathbf{b}_f\right), \tag{26}$$

$$\Gamma_i = \sigma\left(\mathbf{W}_{xi}\mathbf{x}_t + \mathbf{W}_{hi}\mathbf{h}_{t-1} + \mathbf{b}_i\right), \tag{27}$$

$$\Gamma_o = \sigma\left(\mathbf{W}_{xo}\mathbf{x}_t + \mathbf{W}_{ho}\mathbf{h}_{t-1} + \mathbf{b}_o\right), \tag{28}$$

$$\mathbf{c}_t = \Gamma_f \odot \mathbf{c}_{t-1} + \Gamma_i \odot \tilde{\mathbf{c}}_t, \tag{29}$$

$$\mathbf{h}_t = \Gamma_o \odot \tanh\left(\mathbf{c}_t\right), \tag{30}$$

where $\odot$ denotes the Hadamart product, Γ_f denotes the forget gate, Γ_i denotes the input gate, Γ_o denotes the output gate, $\tilde{\mathbf{c}}_t$ denotes the encoding of new information from the observation $\mathbf{x}_t$. Similarly to a vanilla RNN process function, the temporary memory cell $\tilde{\mathbf{c}}_t$ has inputs $\mathbf{h}_{t-1}, \mathbf{x}_t$. Equation (29) shows that, on one hand, the gate Γ_i controls how much information about the new observation encoded in $\tilde{\mathbf{c}}_t$ should be used in the computation of the new cell state $\mathbf{c}_t$. On the other hand, Γ_f controls the amount of past information that should be used in the computation of the new cell state. A key advantage of these models is that the forget and input gates enable an acute control of the gradient flow and thereby prevent gradient vanishing and exploding.

The LSTM was outperformed by other, more complex RNN architectures, for some specific problems [76]. This suggests that even if some architectures might be better than the LSTM, they are not trivial to find and are problem-dependent. Thus, LSTM remains an architecture of choice for approaching a problem involving sequential data.

2 Literature Review

Deep architectures and data-driven learning methods have the potential to substantially improve current medical diagnosis systems [77]. A variety of paths to improve neonatal sepsis detection has been suggested by Celik et al. in [78]. The authors hint toward the shift to more advanced blood sample analysis techniques in order to increase precision and reduce the diagnosis time. New point-of-care (POC) devices for neonatal sepsis diagnosis could enable rapid and accurate bedside assessment of patients, thus providing reliable prognostic information to the medical staff. These methods, although they have been shown promising in animal models, are still being evaluated for potential use in clinical practice [79].

Neonatal sepsis was also shown to alter the expression of gene expression, protein translation, and metabolite production in genome-wide data [80, 81]. This has led to the investigation of multiomics-based prognosis tools for neonatal sepsis. Although promising, the current complexity of retrieving information from multiomics data has led us to study other alternatives for improving neonatal sepsis detection. Neonatal sepsis was shown to affect the ANS, in turn causing clinical physiological symptoms visible on the vital signs, monitored daily in most NICUs.

Patient assessment from physiological information in general, and neonatal sepsis detection in particular, has been investigated in the literature. Static physiological data were shown predictive of mortality in neonates in a large retrospective cohort study [82]. The authors showed that static variables sampled up to 5 minutes after birth could be used to design an artificial neural network prediction score with over 0.9 AUROC performances on a mortality prediction task. Artificial Neural Networks were also shown promising for NSD in another similar study [83]. This study also uses static birth information and reports an AUROC above 0.9. The use of more advanced models such as RNNs for adverse event detection in adult ICU settings has been investigated in the literature. Simple LSTM architectures were shown to have good performances to model irregularly sampled clinical data [84]. The authors also show that LSTM architectures can outperform feature engineering-based baselines when trained on raw hourly-resampled time series data. A similar study was performed in [85], specifically for sepsis blood culture outcome prediction on an adult population. The authors show outstanding performances with over 0.95 AUROC. More advanced methods such as hybrid CNN/RNN were introduced for mortality risk assessment of neonatal patients [86]. The authors use this model on data with a sampling period of 12 hours. Another study also addresses continuous mortality prediction from demographics/static and vital signs data [87]. The authors use the last 80 hours of raw 1 Hz data, segmented into non-overlapping 5 minutes frames (i.e., frames of 1500 samples). The authors implement an interesting model using CNN and LSTMs. Advanced methods based upon Gaussian processes and RNNs have been designed to deal with the irregularly spaced time series data for sepsis predictions in adult [88]. The authors show that their approach outperformed the baseline warning score in place in the ICU. Another approach combining RNNs with a proportional hazard model showed promising results against LR, simple feed-forward neural networks, and simpler hazard models [89]. Studies on adults have mostly been performed on the MIMIC-III dataset, which is an open, public, curated adult ICU dataset. Works implementing RNNs for prediction tasks on the same dataset were shown to improve state of the art baselines, e.g. [90].

The literature for neonatal sepsis detection with recurrent neural networks is more scarce than for adult sepsis. We conjecture that this could be due to the lack of standardized definitions of sepsis for infants and the lack of open databases [64]. We nonetheless find some research items tackling neonatal sepsis detection from ICU data using RNNs. In [91], the authors introduced a model based upon LSTM that showed good generalization abilities across datasets. In [92], an architecture constructed as a combination of LSTM and gated recurrent units [93] (GRU) was used. The authors show that their approach has promising results for neonatal sepsis detection tasks. The authors report an AUROC culminating at 0.9 up to 6 hours prior to sepsis. However, the numerical results are not directly comparable with ours, since (1) the authors use a case–control study setup, enabling them to report AUROC in windows within the "label 1" period, and (2) the authors use a longer septic frame labeling period of 72 hours. In our case, the frames are labeled as septic up to 24 hours, and we use a retrospective cohort study setup. Despite major methodological differences, the numerical results cannot be compared directly with ours, but this study is of particular interest to us although.

3 Experimental Setup for Model Comparison

In this section, we describe the experimental setup we used to evaluate deep recurrent architectures for neonatal sepsis detection in our study [94]. The setup presented in this section was also utilized to assess the performances of convolutional layer-based deep learning architectures in [95].

3.1 Dataset Preparation

We utilized the Philips IntelliVue MX800 Patient Monitor (Philips Healthcare, Amsterdam, Netherlands) to collect high-frequency vital signs data. The time series of labels were constructed based on a clinical event timeline extracted from the Electronic Health Records (EHR) at Karolinska University Hospital, Stockholm, Sweden. We identified the time of blood culture collection as the time of sepsis suspicion. The pre-processed monitoring data, which were resampled at 1 Hz, served as the basis signals. Specifically, we employed the electrocardiogram-derived IBI signal, the level of blood oxygen saturation measured by pulse oximetry (SpO2), and the respiratory frequency derived from chest impedance. To utilize the sample entropy feature[65], we filtered the IBI signal to remove ectopic beats and other strong non-physiological frequency content. This was achieved through a moving median filter of width 3 samples, which was composed with a Butterworth filter of order 6 with cut frequencies of 0.0021 and 0.43 Hz [55, 96].

We extracted features from time frames of duration 55 minutes with 50% overlap, which we found to perform well in a preliminary experimental study [97]. We labeled time frames starting up to 24 hours prior to sepsis diagnosis as $y = 1$, while those starting earlier or directly after sepsis suspicion were labeled as $y = 0$. Time frames containing signals between two sepsis diagnoses, occurring no more than 14 days apart, were labeled as $y = 1$. To characterize the distribution of the signal in a time frame, we calculated the minimum, maximum, mean, standard deviation, skewness, and kurtosis of all signals, as described in Sect. 1.2. Skewness indicates the deviation of the signal distribution from the symmetry of a Gaussian distribution, while kurtosis captures the dispersion of the variance around its expectation [63]. In addition to these features, we computed the sample asymmetry and sample entropy of the IBI signal. These features, along with the standard deviation of the IBI signal, are part of the commercially available HeRO system [62]. We also included sex and birth weight as static demographic features, as well as postnatal age and repetitive body weight measurements.

Overall, our dataset comprises $N = 118$ patients. We denote the dataset as $\mathcal{D} = \{(\mathbf{X}^{(k)}, \mathbf{Y}^{(k)})\}_{k=1}^{N}$, where $\mathbf{X}^{(k)} = [\mathbf{x}_1^{(k)}, \ldots, \mathbf{x}_{T_k}^{(k)}]$ is the series of length T_k of $d = 24$ dimensional samples for patient k and $\mathbf{Y}^{(k)} = [y_1^{(k)}, \ldots, y_{T_k}^{(k)}]$ is the series of length T_k of binary labels for patient k.

The number of patients per population, the frame count, and the prevalence of positive frames are detailed in Table 3. The frame counts are presented as average $\pm$ standard deviation per patient.

3.2 Baseline Models

In this study, we conducted a comparison between the vanilla RNN architecture and the LSTM architecture, with a logistic regression (LR) model serving as the

Table 3 Patient cohort description. The prevalence corresponds to the ratio: $\frac{\#_1}{\#_2}$

		Frames per patient	Sample count	
Population	Patient count	Mean (std)	Total	Prevalence
Overall	118	1 141 (1 027)	134 668	0.48%
Positive	10	2 099 (1 280)	20 992	2.86%
Negative	108	1 053 (952)	113 676	0%

baseline. The LR model treated consecutive window-based feature samples as independent and was optimized with the LBFGS solver on the binary cross-entropy loss, using the sklearn library [98].

For the RNNs, the trainable weights for the linear transforms in the process and measurement functions were initialized from a uniform distribution with support $\mathcal{U}(k) = [-\frac{1}{\sqrt{k}}; \frac{1}{\sqrt{k}}]$, where k is the number of input features to the transform [99]. The dataset used in this study was imbalanced, with approximately 11 times as many patients with label $y = 0$ as patients with label $y = 1$. This presents a challenge, as prediction models trained on such data tend to be biased toward predicting the majority class. To address this issue, we learned our RNN architectures with back-propagation on a weighted cross-entropy loss, where the weights were chosen based on the approach described in Sect. 1.2.

Specifically, for a given patient with a sequence of true labels $\mathbf{Y} \in \mathbb{R}^T$ and corresponding estimated sepsis probabilities $\hat{\mathbf{Y}} \in \mathbb{R}^T$, the weighted cross-entropy loss was calculated using Eq. (31):

$$\mathcal{L}(\mathbf{Y}, \hat{\mathbf{Y}}) = -\sum_{t=1}^{T} w_{y_t} \left(y_t \log(\hat{y}_t) + (1 - y_t) \log(1 - \hat{y}_t) \right), \tag{31}$$

where y_t is the true label at time t, $\hat{y}t$ is the estimated sepsis probability at time t, and w_{y_t} is the weight assigned to the true label y_t.

In this study, we used the Adam optimizer [100], with a fixed learning rate of 0.001 and a fixed number of training epochs of 200. These values were selected based on experimental evaluation and demonstrated a favorable bias-variance trade-off. We fed all the subsequences from a given patient to the RNN algorithms in a single batch. Consequently, the batch size varied between 102 and 4065 subsequences, depending on the input patient. The models were implemented in Python, using the Pytorch library [101].

The classification scores used in this study are discussed in detail in Sect. 1.2. To evaluate our models, we utilized metrics based on the number of true positives (TP), true negatives (TN), false positives (FP), and false negatives (FN). These four basic metrics were used to compute and report the following classification scores: precision $\frac{TP}{TP+FP}$, recall/sensitivity (sen): $\frac{TP}{TP+FN}$, specificity (spec): $\frac{TN}{TN+FP}$, F_1-score: $2\frac{p \times r}{p+r}$, and balanced accuracy (bAcc): $\frac{1}{2}$(sen + spec). In our analysis, individual samples were classified as "positive" when the output probability

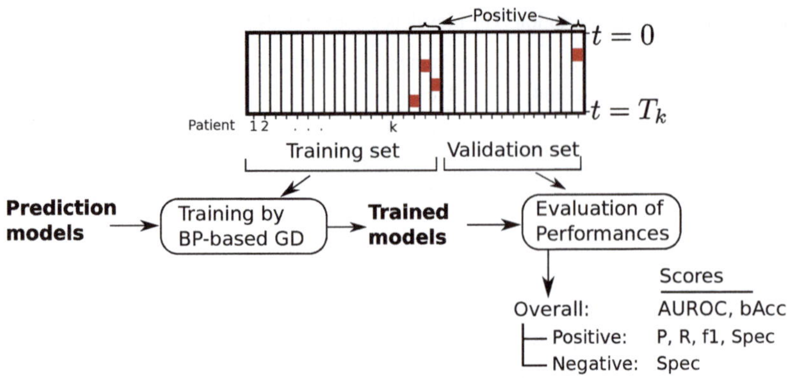

Fig. 8 Training and validation procedure flowchart

$\hat{y}_t > 0.5$. Additionally, we report the threshold-independent area-under-the-receiver-operating-characteristic (AUROC) on the overall population. To validate the generalizability of our methods and make use of as much information as possible from our dataset, we employed a modified leave-one-patient-out cross-validation approach. The process is illustrated in Fig. 8.

The training data consist of all positive patients except one and a random sample of 10% of negative patients, while the remaining 90% of negative patients and the excluded positive patient form the validation data. This strategy ensures that the test data are representative of the population in terms of the prevalence of positive patients. The process is repeated ten times, with each positive patient assigned to the validation set once, until all positive patients have been used. The results are reported as the mean and standard deviation of performance metrics calculated on the validation data. We compared the models' performances on (1) positive patients only, (2) negative patients only, and (3) the entire validation population. By separating patients in the validation data, we were able to evaluate the model's specificity and determine whether it produced more false alarms on positive patients, who are typically the most critically ill.

4 Results and Interpretation

4.1 Numerical Results

The results for our baseline models are presented in Table 4.

The results for LSTM-based models with hidden sizes $h = 50, 100, 150$ and 200, and with a number of recurrent units composing the model of 3 and 4 are reported in Table 5.

Table 4 Performance metrics (mean ± std) for the baseline LR and vanilla RNN. Spec.: Specificity, Prec.: Precision, bAcc: Balanced accuracy

Population	Positive				Negative	Overall	
Score	Prec.	Recall	F1-score	Spec.	Spec.	AUROC	bAcc
LR	**0.06** ± 0.04	**0.60** ± 0.39	**0.11** ± 0.07	**0.73** ± 0.28	**0.77** ± 0.04	**0.81** ± 0.15	0.60 ± 0.23
Vanilla RNN	0.04 ± 0.04	0.59 ± 0.41	0.07 ± 0.07	0.58 ± 0.24	0.70 ± 0.11	0.71 ± 0.18	**0.62** ± 0.18

Table 5 Performance metrics (mean ± std) for various architectures of LSTM-based RNNs

Population		Positive				Negative	Overall	
Hyper-parameter		Scores						
Hidden size	# Layers	Prec.	Recall	F1-score	Spec.	Spec.	AUROC	bAcc
50	3	0.06 ± 0.08	0.23 ± 0.36	0.08 ± 0.12	0.92 ± 0.14	**0.99** ± 0.02	0.72 ± 0.22	0.6 ± 0.18
100	3	**0.17** ± 0.24	0.31 ± 0.39	**0.18** ± 0.26	0.91 ± 0.21	0.97 ± 0.03	0.74 ± 0.26	0.64 ± 0.2
150	3	0.05 ± 0.06	0.19 ± 0.32	0.07 ± 0.09	0.93 ± 0.17	0.97 ± 0.03	0.77 ± 0.23	0.59 ± 0.16
200	3	0.16 ± 0.31	**0.36** ± 0.40	**0.18** ± 0.31	0.87 ± 0.17	0.98 ± 0.03	**0.81** ± 0.18	**0.66** ± 0.2
50	4	0.04 ± 0.12	0.03 ± 0.07	0.03 ± 0.09	**0.94** ± 0.13	0.97 ± 0.03	0.61 ± 0.23	0.5 ± 0.04
100	4	0.12 ± 0.24	0.18 ± 0.33	0.14 ± 0.27	0.91 ± 0.13	0.98 ± 0.02	0.73 ± 0.25	0.58 ± 0.17
150	4	0.07 ± 0.09	0.29 ± 0.41	0.08 ± 0.12	0.9 ± 0.17	0.98 ± 0.04	0.78 ± 0.21	0.63 ± 0.19
200	4	0.11 ± 0.16	0.32 ± 0.43	0.14 ± 0.2	0.93 ± 0.17	0.98 ± 0.02	0.79 ± 0.22	0.64 ± 0.21

The LR model outperformed the vanilla RNN model in terms of AUROC by 14% for the overall population, with AUROC scores of 0.81 and 0.71, respectively. Although the balanced accuracy score was higher for the vanilla RNN model on this population, caution is advised due to the large standard deviation. In contrast, the LR model demonstrated higher specificity (i.e., 0.77) compared to the vanilla RNN (i.e., 0.70) when considering only the negative patients. For positive patients, the specificity of the LR model was 26% higher (i.e. 0.73) than that of the vanilla RNN (i.e. 0.58), although the standard deviation was large. With respect to other metrics, precision, recall, and f1-score, the performances of the two models were similar.

These results suggest that the LR model outperforms the vanilla RNN model by a small margin, indicating that the latter is insufficient in capturing time dependencies in consecutive window-based features. Results are reported for models with varying numbers of layers and hidden sizes. The best performing LSTM-based RNN on the overall population in terms of both AUROC and balanced accuracy (bAcc) is a model with 3 hidden layers and a hidden size of $h = 200$. However, the difference in performances between this model and a similar model with 4 hidden layers is only marginal, given the large standard deviation. All models exhibit similar performances on the negative population, with specificity values of at least 0.97 for all models, and reaching 0.99 for the smallest model with 3 layers and a hidden size of $h = 50$. Across all LSTM-based RNNs, the mean specificity on the positive population is consistently lower, and the standard deviation is higher than that of the negative population. The best model in terms of mean and standard deviation for this metric is an LSTM-based RNN with 4 layers and hidden size 50. This outcome suggests that the models tend to generate less false alarms for negative patients than for positive patients.

The LSTM-based RNN with 3 layers and a hidden size of 200 exhibits the best recall and f1-score, as well as the second-best precision among all models. The model with 3 layers and a hidden size of 100 demonstrates similar performance across all metrics. In contrast, models with 4 hidden layers have lower precision, recall, and f1-score. Considering that models with 3 layers have fewer trainable parameters, we compare these findings to the baseline models that use LSTM-based RNNs with hidden sizes of $h = 100$ and $h = 200$.

Compared to the baseline models, LSTM-based RNNs exhibit a specificity that is +26% higher than the LR model on the negative population and +19% higher on the positive population. Moreover, the standard deviation is lower for both LSTM-based RNNs. However, this improvement in specificity comes at the expense of a 40% lower recall for the LSTM-based RNNs, while their precision remains higher than that of the LR model, at 0.16 versus 0.06. These results suggest that the LSTM-based RNNs are more cautious than the LR models, thus triggering fewer false alarms while missing some positive frames.

4.2 Example Cases

The predicted sequences of sepsis probabilities $\hat{\mathbf{Y}}$ versus postnatal age for two example patients, as well as their corresponding confusion matrices, are presented in Fig. 9.

The estimated sepsis probabilities sequences are generated using LSTM-based RNNs with 3 layers and hidden sizes of $h = 100$ for patient 1 and $h = 200$ for patient 2.

Regarding patient 1, it can be observed that the two peaks in sepsis probability occur slightly after the corresponding labeled segments. This observation is supported by the confusion matrices where very few samples labeled as "1" are accurately predicted as such by the LSTM models. Additionally, it is worth noting that both LSTM-based models exhibit a high sepsis probability between 200 and 350 hours after birth, which falls between two instances of diagnosed sepsis. It

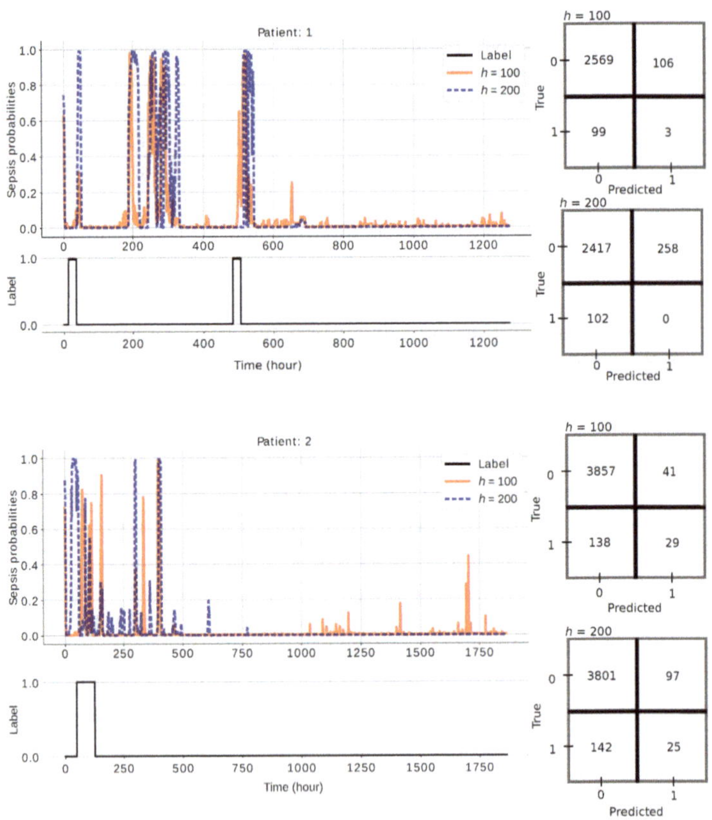

Fig. 9 Time series of predicted sepsis probabilities and associated confusion matrices for two example patients and for two LSTM-based RNN with 3 layers and with hidden size $h = 100, 200$

is possible that our models detect adverse patterns in the vital signs that are not captured by our labeling approach for "septic" time periods.

In the case of patient 2, the LSTM-based models successfully detected the frame within the labeled segment 1. However, the model with a hidden size of 200 produced high sepsis probabilities even before this segment. The confusion matrices reveal that both models incorrectly classified frames within the labeled segment 1, while falsely outputting high probabilities 13–14 days after the positive label segment. Our detailed clinical timeline shows that the patient was unstable until $t = 400$ hours, with multiple reports of apnea and bradycardia. The correctness of our LSTM-based RNN models in labeling 0 segments after the last sepsis diagnosis is supported by the confusion matrices, which indicate that a vast majority of frames are accurately classified as 0 for both patients.

The prediction time series can be computed continuously at bedside. In the setup we discussed, the input data time series predictions can be computed every 22.5 minutes, since the input data are time series of features computed on 55 minutes time frames with 50% overlap. More frequent predictions may be obtained at bedside by retraining the model on consecutive time frames with increased overlap.

5 Conclusion and Future Perspectives

Deep architectures show great potential for neonatal sepsis detection, which is a leading cause of morbidity and mortality in infants. The use of deep learning for analyzing large amounts of medical data has the potential to produce more accurate and timely diagnosis of sepsis in newborns, potentially leading to better clinical decision-making and improved patient outcomes. In the study under scrutiny, we showed that LSTM-based RNNs are both (1) more conservative and (2) more precise than LR or vanilla RNN. LSTM-based RNNs achieved population scores similar to the logistic regression model. Moreover, the increased specificity of the RNN-based models for non-septic patients makes them less prone to false alarms, reducing the risk for alarm fatigue in the clinic. This is a crucial characteristic enabling the deployment of predictive models in clinical practice. We also recognize that the overall performances in terms of precision and recall on the septic patients remain low, while the standard deviation is large. This may be due to poor convergence of the models, inacurate labels, or a too small cohort for training and validation. Thus, further investigations on larger patient populations from various clinical centers are needed to refine and validate these models, as well as to explore their integration with clinical practice. With continued advancements in the understanding of predictive modeling and the availability of large datasets, the future of neonatal sepsis prediction with deep learning looks promising and may ultimately lead to better health outcomes for newborns.

References

1. F. Kim, R. A. Polin, and T. A. Hooven, "Neonatal sepsis," *BMJ*, vol. 371, p. m3672, Oct. 2020.
2. A. Värri, A. Kallonen, E. Helander, A. Ledesma, and P. Pladys, "The Digi-NewB project for preterm infant sepsis risk and maturity analysis," *Finnish Journal of eHealth and eWelfare*, vol. 10, pp. 330–333, May 2018.
3. M. S. Harrison and R. L. Goldenberg, "Global burden of prematurity," *Seminars in Fetal and Neonatal Medicine*, vol. 21, pp. 74–79, Apr. 2016.
4. W. Bank, "New World Bank country classifications by income level: 2021–2022," 2022.
5. H. Blencowe, S. Cousens, M. Z. Oestergaard, D. Chou, A.-B. Moller, R. Narwal, A. Adler, C. Vera Garcia, S. Rohde, L. Say, and J. E. Lawn, "National, regional, and worldwide estimates of preterm birth rates in the year 2010 with time trends since 1990 for selected countries: A systematic analysis and implications," *Lancet*, vol. 379, pp. 2162–2172, June 2012.
6. S. Beck, D. Wojdyla, L. Say, A. Pilar Bertran, M. Meraldi, J. Harris Requejo, C. Rubens, R. Menon, and P. Van Look, "The worldwide incidence of preterm birth: A systematic review of maternal mortality and morbidity," *Bull World Health Org*, vol. 88, pp. 31–38, Jan. 2010.
7. H. Blencowe, S. Cousens, D. Chou, M. Oestergaard, L. Say, A.-B. Moller, M. Kinney, J. Lawn, and Born Too Soon Preterm Birth Action Group, "Born too soon: The global epidemiology of 15 million preterm births," *Reprod Health*, vol. 10 Suppl 1, p. S2, 2013.
8. SNQ, "Årsrapporter," tech. rep., 2020.
9. D. L. Eckberg, "Human sinus arrhythmia as an index of vagal cardiac outflow," *Journal of Applied Physiology*, vol. 54, pp. 961–966, Apr. 1983.
10. S. B. Mulkey, R. B. Govindan, L. Hitchings, T. Al-Shargabi, N. Herrera, C. B. Swisher, A. Eze, S. Russo, S. D. Schlatterer, M. B. Jacobs, R. McCarter, A. Kline, G. L. Maxwell, R. Baker, and A. J. du Plessis, "Autonomic nervous system maturation in the premature extrauterine milieu," *Pediatr Res*, vol. 89, pp. 863–868, Mar. 2021.
11. B. A. Sullivan and K. D. Fairchild, "Vital signs as physiomarkers of neonatal sepsis," *Pediatr Res*, pp. 1–10, Sept. 2021.
12. S. B. Mulkey, S. Kota, C. B. Swisher, L. Hitchings, M. Metzler, Y. Wang, G. L. Maxwell, R. Baker, A. J. du Plessis, and R. Govindan, "Autonomic nervous system depression at term in neurologically normal premature infants," *Early Hum Dev*, vol. 123, pp. 11–16, Aug. 2018.
13. S. R. Yiallourou, N. B. Witcombe, S. A. Sands, A. M. Walker, and R. S. C. Horne, "The development of autonomic cardiovascular control is altered by preterm birth," *Early Hum Dev*, vol. 89, pp. 145–152, Mar. 2013.
14. G. Nino, R. B. Govindan, T. Al-Shargabi, M. Metzler, A. N. Massaro, G. F. Perez, R. McCarter, C. E. Hunt, and A. J. du Plessis, "Premature Infants Rehospitalized because of an Apparent Life-Threatening Event Had Distinctive Autonomic Developmental Trajectories," *Am J Respir Crit Care Med*, vol. 194, pp. 379–381, Aug. 2016.
15. T. Al-Shargabi, D. Reich, R. B. Govindan, S. Shankar, M. Metzler, C. Cristante, R. McCarter, A. D. Sandler, M. Said, and A. du Plessis, "Changes in Autonomic Tone in Premature Infants Developing Necrotizing Enterocolitis," *Am J Perinatol*, vol. 35, pp. 1079–1086, Sept. 2018.
16. V. Tuzcu, S. Nas, U. Ulusar, A. Ugur, and J. R. Kaiser, "Altered Heart Rhythm Dynamics in Very Low Birth Weight Infants With Impending Intraventricular Hemorrhage," *Pediatrics*, vol. 123, pp. 810–815, Mar. 2009.
17. K. D. Fairchild and T. M. O'Shea, "Heart rate characteristics: Physiomarkers for detection of late-onset neonatal sepsis," *Clin Perinatol*, vol. 37, pp. 581–598, Sept. 2010.
18. J. L. Wynn, "Defining neonatal sepsis," *Curr Opin Pediatr*, vol. 28, pp. 135–140, Apr. 2016.
19. K. A. Simonsen, A. L. Anderson-Berry, S. F. Delair, and H. D. Davies, "Early-onset neonatal sepsis," *Clin Microbiol Rev*, vol. 27, pp. 21–47, Jan. 2014.

20. C. Geffers, S. Baerwolff, F. Schwab, and P. Gastmeier, "Incidence of healthcare-associated infections in high-risk neonates: Results from the German surveillance system for very-low-birthweight infants," *Journal of Hospital Infection*, vol. 68, pp. 214–221, Mar. 2008.
21. C. J. Joseph, W. B. Lian, and C. L. Yeo, "Nosocomial Infections (Late Onset Sepsis) in the Neonatal Intensive Care Unit (NICU)," *Proceedings of Singapore Healthcare*, vol. 21, pp. 238–244, Dec. 2012.
22. M. Singh, M. Alsaleem, and C. P. Gray, "Neonatal Sepsis," in *StatPearls*, Treasure Island (FL): StatPearls Publishing, 2022.
23. S. L. Raymond, J. A. Stortz, J. C. Mira, S. D. Larson, J. L. Wynn, and L. L. Moldawer, "Immunological Defects in Neonatal Sepsis and Potential Therapeutic Approaches," *Front Pediatr*, vol. 5, p. 14, 2017.
24. M. K. Van Dyke, C. R. Phares, R. Lynfield, A. R. Thomas, K. E. Arnold, A. S. Craig, J. Mohle-Boetani, K. Gershman, W. Schaffner, S. Petit, S. M. Zansky, C. A. Morin, N. L. Spina, K. Wymore, L. H. Harrison, K. A. Shutt, J. Bareta, S. N. Bulens, E. R. Zell, A. Schuchat, and S. J. Schrag, "Evaluation of universal antenatal screening for group B streptococcus," *N Engl J Med*, vol. 360, pp. 2626–2636, June 2009.
25. M. J. Bizzarro, C. Raskind, R. S. Baltimore, and P. G. Gallagher, "Seventy-five years of neonatal sepsis at Yale: 1928–2003," *Pediatrics*, vol. 116, pp. 595–602, Sept. 2005.
26. C. Fleischmann, F. Reichert, A. Cassini, R. Horner, T. Harder, R. Markwart, M. Tröndle, Y. Savova, N. Kissoon, P. Schlattmann, K. Reinhart, B. Allegranzi, and T. Eckmanns, "Global incidence and mortality of neonatal sepsis: A systematic review and meta-analysis," *Arch Dis Child*, vol. 106, pp. 745–752, Aug. 2021.
27. K. Mayor-Lynn, V. H. González-Quintero, M. J. O'Sullivan, A. I. Hartstein, S. Roger, and M. Tamayo, "Comparison of early-onset neonatal sepsis caused by Escherichia coli and group B Streptococcus," *American Journal of Obstetrics and Gynecology*, vol. 192, pp. 1437–1439, May 2005.
28. J. L. Wynn, S. O. Guthrie, H. R. Wong, P. Lahni, R. Ungaro, M. C. Lopez, H. V. Baker, and L. L. Moldawer, "Postnatal Age Is a Critical Determinant of the Neonatal Host Response to Sepsis," *Mol Med*, vol. 21, pp. 496–504, June 2015.
29. S. Esposito, A. Zampiero, L. Pugni, S. Tabano, C. Pelucchi, B. Ghirardi, L. Terranova, M. Miozzo, F. Mosca, and N. Principi, "Genetic Polymorphisms and Sepsis in Premature Neonates," *PLOS ONE*, vol. 9, p. e101248, July 2014.
30. J. Wynn, T. T. Cornell, H. R. Wong, T. P. Shanley, and D. S. Wheeler, "The Host Response to Sepsis and Developmental Impact," *Pediatrics*, vol. 125, pp. 1031–1041, May 2010.
31. R. Habimana, I. Choi, H. J. Cho, D. Kim, K. Lee, and I. Jeong, "Sepsis-induced cardiac dysfunction: A review of pathophysiology," *Acute Crit Care*, vol. 35, pp. 57–66, May 2020.
32. J. L. Wynn and H. R. Wong, "Pathophysiology and Treatment of Septic Shock in Neonates," *Clin Perinatol*, vol. 37, pp. 439–479, June 2010.
33. A. O. Hofstetter, L. Legnevall, E. Herlenius, and M. Katz-Salamon, "Cardiorespiratory development in extremely preterm infants: Vulnerability to infection and persistence of events beyond term-equivalent age," *Acta Paediatrica*, vol. 97, no. 3, pp. 285–292, 2008.
34. B. Goldstein, B. Giroir, A. Randolph, and International Consensus Conference on Pediatric Sepsis, "International pediatric sepsis consensus conference: Definitions for sepsis and organ dysfunction in pediatrics," *Pediatr Crit Care Med*, vol. 6, pp. 2–8, Jan. 2005.
35. S. Coggins, M. C. Harris, R. Grundmeier, E. Kalb, U. Nawab, and L. Srinivasan, "Performance of Pediatric Systemic Inflammatory Response Syndrome and Organ Dysfunction Criteria in Late-Onset Sepsis in a Quaternary Neonatal Intensive Care Unit: A Case-Control Study," *J Pediatr*, vol. 219, pp. 133–139.e1, Apr. 2020.
36. M. Singer, C. S. Deutschman, C. W. Seymour, M. Shankar-Hari, D. Annane, M. Bauer, R. Bellomo, G. R. Bernard, J.-D. Chiche, C. M. Coopersmith, R. S. Hotchkiss, M. M. Levy, J. C. Marshall, G. S. Martin, S. M. Opal, G. D. Rubenfeld, T. van der Poll, J.-L. Vincent, and D. C. Angus, "The Third International Consensus Definitions for Sepsis and Septic Shock (Sepsis-3)," *JAMA*, vol. 315, pp. 801–810, Feb. 2016.

37. T. Aşuroğlu and H. Oğul, "A deep learning approach for sepsis monitoring via severity score estimation," *Computer Methods and Programs in Biomedicine*, vol. 198, p. 105816, Jan. 2021.
38. J. L. Wynn and R. A. Polin, "A neonatal sequential organ failure assessment score predicts mortality to late-onset sepsis in preterm very low birth weight infants," *Pediatric Research*, vol. 88, pp. 85–90, July 2020.
39. N. Fleiss, S. A. Coggins, A. N. Lewis, A. Zeigler, K. E. Cooksey, L. A. Walker, A. N. Husain, B. S. de Jong, A. Wallman-Stokes, M. W. Alrifai, D. H. Visser, M. Good, B. Sullivan, R. A. Polin, C. R. Martin, and J. L. Wynn, "Evaluation of the Neonatal Sequential Organ Failure Assessment and Mortality Risk in Preterm Infants With Late-Onset Infection," *JAMA Network Open*, vol. 4, p. e2036518, Feb. 2021.
40. V. Siljehav, A. M. Hofstetter, K. Leifsdottir, and E. Herlenius, "Prostaglandin E2 Mediates Cardiorespiratory Disturbances during Infection in Neonates," *The Journal of Pediatrics*, vol. 167, pp. 1207–1213.e3, Dec. 2015.
41. M. Puia-Dumitrescu, D. T. Tanaka, T. G. Spears, C. J. Daniel, K. R. Kumar, K. Athavale, S. E. Juul, and P. B. Smith, "Patterns of phlebotomy blood loss and transfusions in extremely low birth weight infants," *J Perinatol*, vol. 39, pp. 1670–1675, Dec. 2019.
42. J. A. Widness, "Pathophysiology of Anemia During the Neonatal Period, Including Anemia of Prematurity," *Neoreviews*, vol. 9, p. e520, Nov. 2008.
43. W. Hellström, T. Martinsson, A. Hellstrom, E. Morsing, and D. Ley, "Fetal haemoglobin and bronchopulmonary dysplasia in neonates: An observational study," *Archives of Disease in Childhood - Fetal and Neonatal Edition*, vol. 106, pp. 88–92, Jan. 2021.
44. E. Persad, G. Sibrecht, M. Ringsten, S. Karlelid, O. Romantsik, T. Ulinder, I. J. B. do Nascimento, M. Björklund, A. Arno, and M. Bruschettini, "Interventions to minimize blood loss in very preterm infants—A systematic review and meta-analysis," *PLOS ONE*, vol. 16, p. e0246353, Feb. 2021.
45. B. Burstein, M. Beltempo, and P. S. Fontela, "Role of C-Reactive Protein for Late-Onset Neonatal Sepsis," *JAMA Pediatrics*, vol. 175, pp. 101–102, Jan. 2021.
46. V. E. Visser and R. T. Hall, "Urine culture in the evaluation of suspected neonatal sepsis," *J Pediatr*, vol. 94, pp. 635–638, Apr. 1979.
47. C. Gabay and I. Kushner, "Acute-phase proteins and other systemic responses to inflammation," *N Engl J Med*, vol. 340, pp. 448–454, Feb. 1999.
48. B. A. Sullivan, V. P. Nagraj, K. L. Berry, N. Fleiss, A. Rambhia, R. Kumar, A. Wallman-Stokes, Z. A. Vesoulis, R. Sahni, S. Ratcliffe, D. E. Lake, J. R. Moorman, and K. D. Fairchild, "Clinical and vital sign changes associated with late-onset sepsis in very low birth weight infants at 3 NICUs," *J Neonatal Perinatal Med*, vol. 14, no. 4, pp. 553–561, 2021.
49. P. B. Batchelder and D. M. Raley, "Maximizing the laboratory setting for testing devices and understanding statistical output in pulse oximetry," *Anesth Analg*, vol. 105, pp. S85–S94, Dec. 2007.
50. A. A. Alian and K. H. Shelley, "Photoplethysmography," *Best Practice & Research Clinical Anaesthesiology*, vol. 28, pp. 395–406, Dec. 2014.
51. A. Hertzman and C. Spealman, "Observations on the finger volume pulse recorded photoelectrically," *Am. J. Physiol.*, vol. 119, pp. 334–335, 1937.
52. M. Nitzan, A. Romem, and R. Koppel, "Pulse oximetry: Fundamentals and technology update," *Med Devices (Auckl)*, vol. 7, pp. 231–239, July 2014.
53. A. Jubran, "Pulse oximetry," *Crit Care*, vol. 19, no. 1, p. 272, 2015.
54. T. Tamura, "Current progress of photoplethysmography and SPO2 for health monitoring," *Biomed Eng Lett*, vol. 9, pp. 21–36, Feb. 2019.
55. D. Nabil and F. Bereksi Reguig, "Ectopic beats detection and correction methods: A review," *Biomedical Signal Processing and Control*, vol. 18, pp. 228–244, Apr. 2015.
56. J. E. Pallett and J. W. Scopes, "Recording respirations in newborn babies by measuring impedance of the chest," *Med. Electron. Biol. Engng*, vol. 3, pp. 161–168, Apr. 1965.
57. J. M. Di Fiore, "Neonatal cardiorespiratory monitoring techniques," *Seminars in Neonatology*, vol. 9, pp. 195–203, June 2004.

58. A. V. Sahakian, W. J. Tompkins, and J. G. Webster, “Electrode Motion Artifacts in Electrical Impedance Pneumography,” *IEEE Transactions on Biomedical Engineering*, vol. BME-32, pp. 448–451, June 1985.
59. P. H. Charlton, T. Bonnici, L. Tarassenko, D. A. Clifton, R. Beale, P. J. Watkinson, and J. Alastruey, “An impedance pneumography signal quality index: Design, assessment and application to respiratory rate monitoring,” *Biomedical Signal Processing and Control*, vol. 65, p. 102339, Mar. 2021.
60. J. Moeyersons, J. Morales, N. Seeuws, C. Van Hoof, E. Hermeling, W. Groenendaal, R. Willems, S. Van Huffel, and C. Varon, “Artefact Detection in Impedance Pneumography Signals: A Machine Learning Approach,” *Sensors*, vol. 21, p. 2613, Jan. 2021.
61. E. Helander, N. Khodor, A. Kallonen, A. Värri, H. Patural, G. Carrault, and P. Pladys, “Comparison of linear and non-linear heart rate variability indices between preterm infants at their theoretical term age and full term newborns,” in *EMBEC & NBC 2017* (H. Eskola, O. Väisänen, J. Viik, and J. Hyttinen, eds.), vol. 65, pp. 153–156, Singapore: Springer Singapore, 2018.
62. J. F. Hicks and K. Fairchild, “Heart rate observation (HeRO) monitoring was developed for detection of sepsis in preterm infants.[…] The HeRO monitor is now in use in many NICUs in the USA and was approved in 2012 for use in Europe.,” p. 5, 2013.
63. R. Joshi, D. Kommers, L. Oosterwijk, L. Feijs, C. Van Pul, and P. Andriessen, “Predicting Neonatal Sepsis Using Features of Heart Rate Variability, Respiratory Characteristics and ECG-Derived Estimates of Infant Motion,” *IEEE Journal of Biomedical and Health Informatics*, pp. 1–1, 2019.
64. E. Persad, K. Jost, A. Honoré, D. Forsberg, K. Coste, H. Olsson, S. Rautiainen, and E. Herlenius, “Neonatal sepsis prediction through clinical decision support algorithms: A systematic review,” *Acta Paediatrica*, vol. 110, no. 12, pp. 3201–3226, 2021.
65. D. E. Lake, J. S. Richman, M. P. Griffin, and J. R. Moorman, “Sample entropy analysis of neonatal heart rate variability,” *American Journal of Physiology-Regulatory, Integrative and Comparative Physiology*, vol. 283, pp. R789–R797, Sept. 2002.
66. S. M. Pincus, “Approximate entropy as a measure of system complexity.,” *Proceedings of the National Academy of Sciences*, vol. 88, pp. 2297–2301, Mar. 1991.
67. B. P. Kovatchev, L. S. Farhy, H. Cao, M. P. Griffin, D. E. Lake, and J. R. Moorman, “Sample Asymmetry Analysis of Heart Rate Characteristics with Application to Neonatal Sepsis and Systemic Inflammatory Response Syndrome,” *Pediatr Res*, vol. 54, pp. 892–898, Dec. 2003.
68. M. Stone, “Cross-Validatory Choice and Assessment of Statistical Predictions,” *Journal of the Royal Statistical Society: Series B (Methodological)*, vol. 36, no. 2, pp. 111–133, 1974.
69. A. Niculescu-Mizil and R. Caruana, “Predicting good probabilities with supervised learning,” in *Proceedings of the 22nd International Conference on Machine Learning - ICML ’05*, (Bonn, Germany), pp. 625–632, ACM Press, 2005.
70. J. P. Egan, *Signal Detection Theory and ROC-analysis*. Academic press, 1975.
71. T. Fawcett, “An introduction to ROC analysis,” *Pattern Recognition Letters*, vol. 27, pp. 861–874, June 2006.
72. C. X. Ling and V. S. Sheng, “Cost-Sensitive Learning,” in *Encyclopedia of Machine Learning* (C. Sammut and G. I. Webb, eds.), pp. 231–235, Boston, MA: Springer US, 2010.
73. D. E. Rumelhart, G. E. Hinton, and R. J. Williams, “Learning representations by back-propagating errors,” *Nature*, vol. 323, pp. 533–536, Oct. 1986.
74. R. Pascanu, T. Mikolov, and Y. Bengio, “On the difficulty of training recurrent neural networks,” in *Proceedings of the 30th International Conference on Machine Learning*, pp. 1310–1318, PMLR, May 2013.
75. S. Hochreiter and J. Schmidhuber, “Long Short-Term Memory,” *Neural Computation*, vol. 9, pp. 1735–1780, Nov. 1997.
76. R. Jozefowicz, W. Zaremba, and I. Sutskever, “An Empirical Exploration of Recurrent Network Architectures,” in *Proceedings of the 32nd International Conference on Machine Learning*, pp. 2342–2350, PMLR, June 2015.

77. R. Vinuesa, H. Azizpour, I. Leite, M. Balaam, V. Dignum, S. Domisch, A. Felländer, S. D. Langhans, M. Tegmark, and F. Fuso Nerini, "The role of artificial intelligence in achieving the Sustainable Development Goals," *Nat Commun*, vol. 11, p. 233, Jan. 2020.
78. I. H. Celik, M. Hanna, F. E. Canpolat, and Mohan Pammi, "Diagnosis of neonatal sepsis: The past, present and future," *Pediatr Res*, vol. 91, pp. 337–350, Jan. 2022.
79. T. Oeschger, D. McCloskey, V. Kopparthy, A. Singh, and D. Erickson, "Point of care technologies for sepsis diagnosis and treatment," *Lab Chip*, vol. 19, pp. 728–737, Feb. 2019.
80. M. Abbas and Y. EL-Manzalawy, "Machine learning based refined differential gene expression analysis of pediatric sepsis," *BMC Medical Genomics*, vol. 13, p. 122, Aug. 2020.
81. P.-Y. Iroh Tam and C. M. Bendel, "Diagnostics for neonatal sepsis: Current approaches and future directions," *Pediatr Res*, vol. 82, pp. 574–583, Oct. 2017.
82. M. Podda, D. Bacciu, A. Micheli, R. Bellù, G. Placidi, and L. Gagliardi, "A machine learning approach to estimating preterm infants survival: Development of the Preterm Infants Survival Assessment (PISA) predictor," *Scientific Reports*, vol. 8, p. 13743, Sept. 2018.
83. F. López-Martínez, E. R. Núñez-Valdez, J. Lorduy Gomez, and V. García-Díaz, "A neural network approach to predict early neonatal sepsis," *Computers & Electrical Engineering*, vol. 76, pp. 379–388, June 2019.
84. Z. C. Lipton, D. C. Kale, C. Elkan, and R. Wetzel, "Learning to Diagnose with LSTM Recurrent Neural Networks," Mar. 2017.
85. T. Van Steenkiste, J. Ruyssinck, L. De Baets, J. Decruyenaere, F. De Turck, F. Ongenae, and T. Dhaene, "Accurate prediction of blood culture outcome in the intensive care unit using long short-term memory neural networks," *Artificial Intelligence in Medicine*, vol. 97, pp. 38–43, June 2019.
86. S. Baker, W. Xiang, and I. Atkinson, "Hybridized neural networks for non-invasive and continuous mortality risk assessment in neonates," *Computers in Biology and Medicine*, vol. 134, p. 104521, July 2021.
87. J. Feng, J. Lee, Z. A. Vesoulis, and F. Li, "Predicting mortality risk for preterm infants using deep learning models with time-series vital sign data," *npj Digital Medicine*, vol. 4, pp. 1–8, July 2021.
88. J. Futoma, S. Hariharan, and K. Heller, "Learning to detect sepsis with a multitask Gaussian process RNN classifier," in *Proceedings of the 34th International Conference on Machine Learning* (D. Precup and Y. W. Teh, eds.), vol. 70 of *Proceedings of Machine Learning Research*, pp. 1174–1182, PMLR, Aug. 2017.
89. S. P. Shashikumar, C. S. Josef, A. Sharma, and S. Nemati, "DeepAISE – An interpretable and recurrent neural survival model for early prediction of sepsis," *Artificial Intelligence in Medicine*, vol. 113, p. 102036, Mar. 2021.
90. M. Scherpf, F. Gräßer, H. Malberg, and S. Zaunseder, "Predicting sepsis with a recurrent neural network using the MIMIC III database," *Computers in Biology and Medicine*, vol. 113, p. 103395, Oct. 2019.
91. R. H. Alvi, M. H. Rahman, A. A. S. Khan, and R. M. Rahman, "Deep learning approach on tabular data to predict early-onset neonatal sepsis," *Journal of Information and Telecommunication*, vol. 5, pp. 226–246, Apr. 2021.
92. C. León, P. Pladys, A. Beuchée, and G. Carrault, "Recurrent Neural Networks for Early Detection of Late Onset Sepsis in Premature Infants Using Heart Rate Variability," in *2021 Computing in Cardiology (CinC)*, vol. 48, pp. 1–4, Sept. 2021.
93. K. Cho, B. van Merrienboer, D. Bahdanau, and Y. Bengio, "On the Properties of Neural Machine Translation: Encoder-Decoder Approaches," *arXiv:1409.1259 [cs, stat]*, Oct. 2014.
94. A. Honoré, H. Siren, R. Vinuesa, S. Chatterjee, and E. Herlenius, "An LSTM-based Recurrent Neural Network for Neonatal Sepsis Detection in Preterm Infants," in *2022 IEEE Signal Processing in Medicine and Biology Symposium (SPMB)*, pp. 1–6, Dec. 2022.
95. H. Siren, "Sequential Deep Learning Models for Neonatal Sepsis Detection," 2022.
96. S. Akselrod, D. Gordon, F. A. Ubel, D. C. Shannon, A. C. Berger, and R. J. Cohen, "Power spectrum analysis of heart rate fluctuation: A quantitative probe of beat-to-beat cardiovascular control," *Science*, vol. 213, pp. 220–222, July 1981.

97. A. Honoré, D. Forsberg, K. Jost, K. Adolphson, A. Stålhammar, E. Herlenius, and S. Chatterjee, "Classification and feature extraction for neonatal sepsis detection," Mar. 2022.
98. F. Pedregosa, G. Varoquaux, A. Gramfort, V. Michel, B. Thirion, O. Grisel, M. Blondel, P. Prettenhofer, R. Weiss, V. Dubourg, J. Vanderplas, A. Passos, and D. Cournapeau, "Scikit-learn: Machine Learning in Python," *MACHINE LEARNING IN PYTHON*, p. 6.
99. K. He, X. Zhang, S. Ren, and J. Sun, "Delving Deep into Rectifiers: Surpassing Human-Level Performance on ImageNet Classification," Feb. 2015.
100. D. P. Kingma and J. Ba, "Adam: A Method for Stochastic Optimization," *arXiv:1412.6980 [cs]*, Jan. 2017.
101. A. Paszke, S. Gross, F. Massa, A. Lerer, J. Bradbury, G. Chanan, T. Killeen, Z. Lin, N. Gimelshein, L. Antiga, A. Desmaison, A. Köpf, E. Yang, Z. DeVito, M. Raison, A. Tejani, S. Chilamkurthy, B. Steiner, L. Fang, J. Bai, and S. Chintala, "PyTorch: An imperative style, high-performance deep learning library," in *Proceedings of the 33rd International Conference on Neural Information Processing Systems*, no. 721 in 1, pp. 8026–8037, Red Hook, NY, USA: Curran Associates Inc., Dec. 2019.

Real-Time ECG Analysis with the ArdMob-ECG: A Comparative Assessment

Tim J. Möller, Moritz Wunderwald, and Markus Tünte

1 Introduction

In recent years, a growing recognition of the need to revolutionize traditional clinical ECG systems, addressing challenges such as limited accessibility and high costs, has been visible. The advent of mobile and portable ECG devices has emerged as a promising solution, with the potential to democratize the access to ECG monitoring. This transformative shift from expensive, unavailable, and bulky, paper-based systems to compact digital devices has been fueled by the integration of advanced sensors, wireless connectivity, and smartphone/tablet applications as well as smartwatches. These technological advancements have paved the way for affordable and high-quality ECG devices capable of real-time analysis. We present the "ArdMob-ECG", which is a novel, portable, easy-to-assemble, and low-cost ECG device with various empirical and clinical, but also cultural, uses, e.g., in live music performances that harnesses the rhythmic cadence of an electrocardiogram tone,

T. J. Möller (✉)
Berlin School of Mind and Brain, Humboldt-Universität zu Berlin, Berlin, Germany

Department for Psychiatry and Psychotherapy, Charité University Medicine, Berlin, Germany

Touro University Berlin, Berlin, Germany
e-mail: tim.julian.moeller@gmail.com

M. Wunderwald
Department of Developmental and Educational Psychology, University of Vienna, Vienna, Austria
e-mail: code@moritzwunderwald.de

M. Tünte
Department of Developmental and Educational Psychology, University of Vienna, Vienna, Austria

Vienna Doctoral School Cognition, Behavior and Neuroscience, University of Vienna, Vienna, Austria
e-mail: markus.tuente@univie.ac.at

A. Ahmed, J. Picone (eds.), *Machine Learning Applications in Medicine and Biology*,
https://doi.org/10.1007/978-3-031-51893-5_6

seamlessly transforming it into a captivating melody with other musical instruments. The ArdMob-ECG has been of use for human–machine interfaces (e.g., virtual and augmented reality settings), remote settings where ECG machines are not available or too expensive (e.g., remote patient monitoring and home-based healthcare), and in settings of limited space (e.g., Saunas and Yoga studios). Detailed technical specifications of the device were introduced in [1] and [2]. Here, a validation to the device's performance as an alternative to medical ECG devices is provided. ECG devices are widely used in science and medicine to monitor heart activity and to detect arrhythmias. However, ECG devices are often expensive, heavy, and require trained medical personnel to operate. Further, ECG devices are often not available in remote areas, hospitalization is not possible, or too expensive. To overcome these limitations, mobile ECGs can be an alternative solution. Specifically, the ArdMob-ECG has been developed, which is a technical device that is easy to use, lightweight, inexpensive, and reliable for measuring heart rate and its components. Another advantage is that the ArdMob-ECG only requires a three-electrode setup instead of a 12-electrode setup, which is usually applied for clinical ECG assessments. The ArdMob-ECG can also be operated via a battery/powerbank or a Laptop, making an energy outlet obsolete and the device completely mobile, with low power consumption of around 25 mA. The ArdMob-ECG integrates the AD8232 Single Lead Heart Rate monitor module (Sparkfun, manufactured by Analog Devices Inc., Wilmington, Massachusetts, United States of America) that is soldered onto an Arduino Mega prototype shield (RobotDyn, Hong Kong, Kwai Cheong rd., NI, China), which can be plugged onto an Arduino Mega board (Arduino, New York, New York, United States). The device also allows for the implementation of an SPI reader to save the data on a microSD-card (AZDelivery, Deggendorf, Germany). It allows to record different data, such as timestamps or raw data, but can also be used for heartbeat detection. The ArdMob-ECG features the "R-Detect", which uses a simplification of the Pan–Tompkins algorithm to detect heartbeats in near real time. Since the algorithm is calculated locally on the ArdMob-ECG, triggers can be sent with minimal delay. The ArdMob-ECG can also play auditory feedback via a Piezo sound buzzer (UKCOCO DC 3–24 V 85 dB Piezo buzzer) whenever a heartbeat is detected, or play these tones faster, or slower, or with a specific delay—a functionality that is not offered by classic clinical machines, but required for several scientific paradigms (e.g. the widely used interoceptive sensitivity paradigm [3–6]. The ArdMob-ECG is shown in Fig. 1, a module connection as well as a schematic diagram is shown under Fig. 2. To enhance signal quality, we employed standard signal processing techniques, including a lowpass filter to eliminate high-frequency noise and artifacts and a highpass filter to attenuate low-frequency noise and baseline wander. Additionally, we utilized a moving average filter from the Pan–Tompkins algorithm to smoothen signals and facilitate the detection of relevant features, such as QRS complexes in ECG signals. Comparatively, the Pan–Tompkins algorithm has proven effective in accurately detecting QRS complexes, offering advantages over alternative methods (i.e. neural networks or Wavelet Transform–based algorithms) that may have increased computational complexity or reduced accuracy in certain signal types [7]. The ArdMob-ECG entails an Arduino

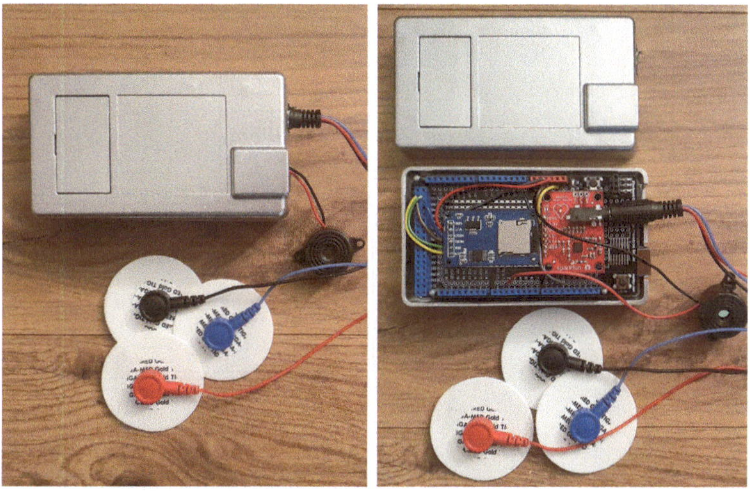

(a) (b)

Fig. 1 (**a**) The ArdMob-ECG device including all optional functionalities. (**b**) The ArdMob-ECG inside its optional aluminium case, visually displaying the sensors and modules as well as the cable connections, according to the schematic in Fig. 2. Pictures taken from [2]

Mega 2560, which has a 10-bit analog-to-digital converter (ADC), providing a signal resolution of 1024 discrete levels. This means it represents analog voltage changes with a granularity of approximately 0.0049 volts. However, with other ADC modules, an even higher granular resolution can be achieved. The AD8232 chip is designed to improve signal resolution in ECG applications. It achieves this through features like low noise amplification, adjustable gain, and high common-mode rejection ratio (CMRR). These enhancements enhance the quality of the ECG waveform by reducing interference and preserving small variations, enabling more precise detection and analysis of important features like QRS complexes. Due to the integration of the three heart derivations into one integrated waveform, some information regarding the diagnostic capabilities of the ECG is however reduced in favor for a clearer R-peak detection [8].

In this chapter, we validated the performance of the ArdMob-ECG using the R-detect function by assessing its usability, validity, and reliability. Furthermore, its delay and the ability to detect heartbeats were assessed. Specifically, the ArdMob-ECG was compared to a state-of-the-art ECG in two use cases. The first use case involved evaluating the accuracy of the R-detect function of the ArdMob-ECG in detecting inter-beat intervals (IBIs), which are important for computing indices such as heart rate variability (HRV) and respiratory sinus arrhythmia (RSA). The second use case involved measuring the latency of the ArdMob-ECG in online detection of R-peaks, which can be used to provide real-time feedback in experimental paradigms such as the heartbeat counting task [9] or the heartbeat detection task [10].

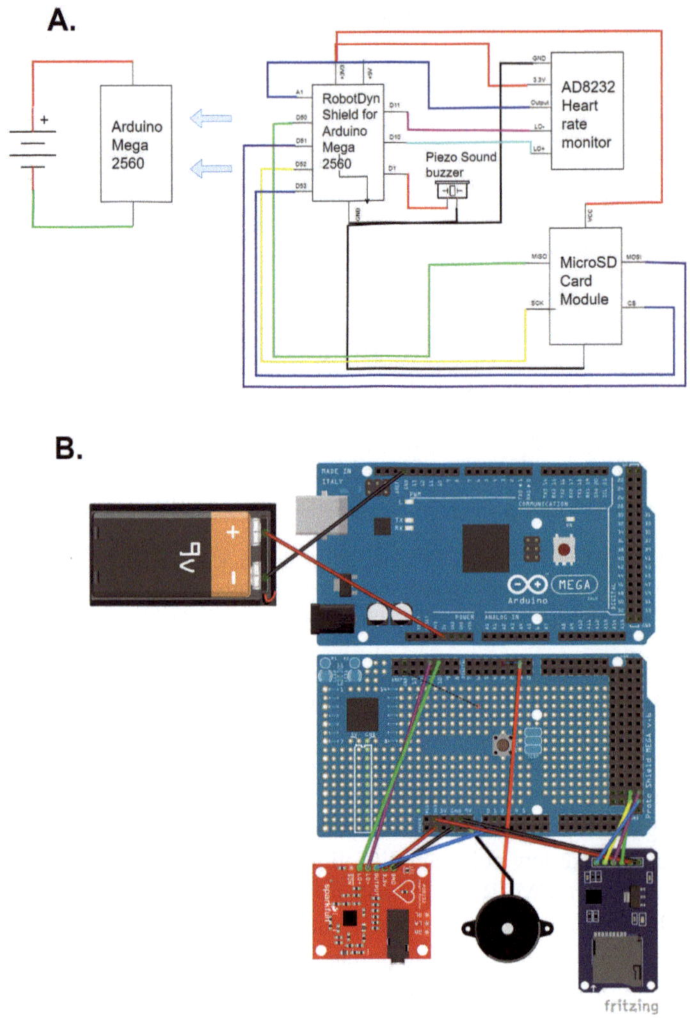

Fig. 2 (**a**) A module connection diagram that shows the connection of the individual modules. (**b**) A graphical depiction of the module connection of the ArdMob-ECG using an Arduino Mega 2560 board (top, teal). The Arduino Mega can be connected to an AD8232 heart beat monitor (bottom left, red) and a 85 dB Piezo sound buzzer (bottom center, black), and a micro SD memory card SPI reader (bottom right, dark blue). It can also be battery-powered (top left). The prototype shield can be attached to the Arduino Mega. Standard shielded cables can be used for the connections. The battery is usually stored outside of the case to reduce electrical interference

1.1 Characterisitcs of an Electrocardiogramm and Pulse Oximetry

An Electrocardiogram (ECG) is a painless and noninvasive technique that can be used to assess heart rate and rhythm. Usually, the QRS complex is in focus when analyzing heartbeat data. It consists of three waves, the Q, R, and S wave. Especially the R-peak is important for scientific use, i.e., to calculate the heart rate variability (HRV). The time between two consecutive R-peaks is called the Interbeat interval (IBI).

Pulse oximetry on the other hand is a noninvasive and painless method to measure oxygen saturation in arterial blood by shining light through a body part, usually a finger, and measuring the amount of light absorbed by oxygenated and deoxygenated hemoglobin. Through that technique, heartbeats can be inferred. The advantages of pulse oximetry are its noninvasiveness, easiness of use, relatively cheap, and easy technological implementation, as well as quick results, making it a valuable tool in clinical and scientific settings. However, pulse oximetry has severe limitations, including inaccurate results in patients with poor perfusion or low hemoglobin levels, high interpersonal variance depending on placement of the sensor, and physiological factors, especially height, weight, arm length, sex, blood pressure, and body fat content [11]. Furthermore, it has been shown that pulse oximetry is less precise in people with darker skin due to differences in light absorption of the skin [12], and that pulse oximetry has a long delay of up to several 100 ms in detecting heartbeats [13]. A pulse oximeter measures the oxygen saturation of arterial blood and pulse rate but does not provide information about the electrical activity of the heart, limiting its utility [14].

Other newly developed methods involve Photoplethysmography (PPG), which utilizes optical sensors in wearable devices to measure blood volume changes in peripheral tissues, enabling real-time heart rate monitoring and with newer technology even the assessment of hypertension [15, 16]. Ballistocardiography (BCG) is another method that detects mechanical body movements caused by cardiac activity, offering continuous heart rate monitoring without direct contact that can also be of use in diagnosing cardiovascular diseases [17]. Impedance cardiography (ICG) measures thoracic impedance changes to estimate stroke volume and cardiac output. For a review and recent applications and developments, see [18]. Lastly, even contactless camera-based remote photoplethysmographic (rPPG) techniques in synergy with modern AI technology by exploiting spatial redundancies of image sensors to distinguish pulse from motion induced signals or by Eulerian Magnification are currently being developed [19–21]. These methods, coupled with advanced signal processing and machine learning techniques, enable accurate and comprehensive cardiovascular monitoring in various settings. Recent investigations, such as of [22] and [23], have assessed the accuracy and reliability of consumer-grade wearable heart rate monitors. In another study by Nunan et al. [24], the authors concluded that spectral measures are extremely sensitive to technical issues within RR-data. Specifically artifacts, misplacement of missing data, poor pre-processing, and nonstationarity are thereby the most problematic sources of error [24]. Especially the application of wavelet transforms [25] and frequency domain analyses [26] for the analysis of ECG data pose a helpful contribution in understanding and interpreting R-R intervals [25], since the deconstruction into its frequency components and time components allows for the identification of different frequency bands and their contribution to the ECG signal. Furthermore, a recent study by Gilgen-Ammann et al. [27] shed light on the dynamical features of the R-R intervals during physical exertion and the accompanying problems in decreasing signal quality.

2 Methods

Five minutes of resting state data from four healthy individuals in the age range between 24 and 30 were collected in separate sessions, two of which were female. During the sessions, we recorded and extracted the ECG or the pulse oximetry from four sources that are listed in Table 1.

In the first case, a three-lead setup from a state-of-the-art ECG (AdInstruments Powerlab 4/35 & BioAmp) setup was used. The data were analyzed using Neurokit2, which is a Python library that provides tools for analyzing physiological and neuroimaging data.

In the second case, a three-lead setup from the ArdMob-ECG utilizing the R-detect function, which, upon detection of a R-peak, sends a trigger directly into the Labchart ADInstruments software (https://www.adinstruments.com/support/downloads/windows/labchart) was used. The third case utilized a finger pulse sensor (AdInstruments). Lastly, a built-in function of the Adinstruments Powerlab (FastResponseOutput), which sends out a pulse upon detection of a R-peak with minimal latency, was used (<1 ms, https://www.adinstruments.com/products/fast-response-output). Data were recorded in Labchart pro, recording with 1000 Hz. In this setup, the devices were connected to a Laptop.

Table 1 Overview for methods used to record ECG signal and identify R-peaks

Method	Practical application	R-peak identification
1. AdInstruments ECG	Three-lead setup (3 electrodes)	Neurokit2 peak identification and manual inspection
2. ArdMob-ECG	Three-lead setup (3 electrodes)	Custom algorithm "R-detect", sending a trigger upon identification of a R-peak
3. AdInstruments Finger Sensor	Finger sensor on right index finger measuring changes in blood pressure	Neurokit2 peak identification manual inspection
4. AdInstruments FastResponseOutput	Three-lead setup (3 electrodes)	Trigger via an analog output send upon detection of input signal (AdInstruments ECG) reaching a predefined threshold

The IBIs for all four signal streams were extracted. For the three-lead setup of the AdInstruments ECG and the finger-pulse sensor, the Neurokit2 was used to identify R-peaks, and any misidentified peaks were manually corrected. For the ArdMob-ECG and the FastResponseOutput, IBIs from the triggers received were calculated. It is important to note, however, that different methods may yield varied waveform characteristics and physiological information; We chose to apply the same outlier measure to provide a standardized approach for detecting and analyzing potential outliers across all signals. However, depending on the pathophysiological conditions, applying different outlier classifiers might be beneficial. Moreover, while time–frequency analysis techniques, such as time–frequency analyses, offer valuable insights into signal characteristics, it is crucial to recognize that there might be better suiting options, depending on the specific method, aims, and application.

Additionally, the actual round trip time (rtt) of a simple ArdMob-ECG to Unity, back to an Arduino setup, was investigated. Unity is a common cross-platform game development engine to allow developers to create interactive and immersive computer programs. To that end, the ArdMob-ECG and an Arduino Uno were both connected to a computer that was running a simple script in Unity 3D. The code can be accessed under https://github.com/wunderwald/Arduino_ECG. The ArdMob-ECG was sending a trigger to Unity, which in turn was sending a ping to the other Arduino, once the trigger was received. This was repeated 27 times in total. A graphical depiction of the setup can be seen in Fig. 4. The data as well as the software code can be found here: https://github.com/wunderwald/Arduino_ECG. The round trip time is a crucial factor in assessing how quick an exchange of information and feedback between two system components can take place. It is essential in assessing the system's responsiveness and determining whether it meets the requirements of real-time processing. Real time refers to processes that can respond to events within a specific timeframe. However, there is no clear definition of real time, and it is highly dependent on the application domain. In case of real-time ECG monitoring (near), real time means that there is a negligible and imperceptible delay to the user in the monitoring and feedback system.

Latencies for (near) real-time feedback can stem from two sources: hardware and software latencies. Software latencies refer to the time that is needed for receiving and processing of the signal in a given software, for instance, running on a presentation computer. Hardware latencies refer to the time that is needed to forward the signal by the ArdMob-ECG. Figure 3 shows a block diagram of the main sources of hardware and software latencies in classic ArdMob-ECG to Computer setups. To investigate the hardware latency R-peaks detected by the AdInstruments, ECG was used as a reference. Next, the time difference in receiving a trigger sent from the ArdMob-ECG upon detection of a R-peak was computed. By comparison, a study by Desai et al. [28] could show that the maximum rate of which heart beat data could be supplied to a system with a clock of 50 MHZ was with a latency of 0.82 seconds. One factor that further affects the time delay is the window size of the algorithm; In this context, the window size refers to the number of consecutive data points that are used to process if an R-peak occurred. Having a bigger windowsize with more datapoints might increase the sensitivity of the algorithm, but at the

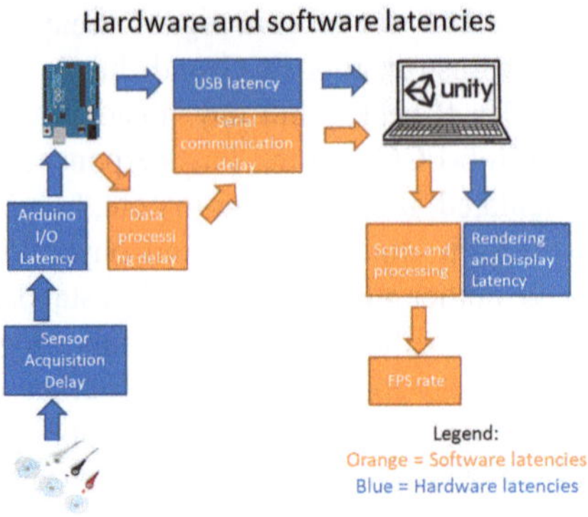

Fig. 3 A block diagram of the main sources of hardware (blue) and software (orange) latencies in a classic ArdMob-ECG to Computer interface

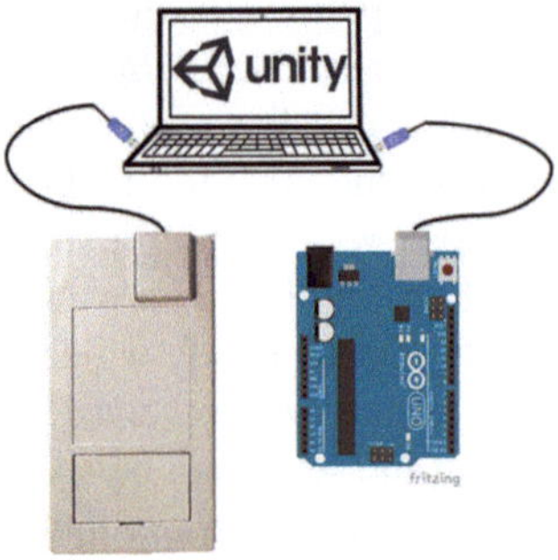

Fig. 4 A depiction of the ArdMob-Unity-Arduino setup. The ArdMob-ECG and another Arduino Uno were connected to a computer running Unity 3D. The ArdMob-ECG was sending a trigger via its serial interface, the computer using a simple Unity program processed the trigger upon arrival and was then sending another trigger to the Arduino Uno. The time necessary for this process is referred to as roundtrip time (rtt). Figure created using Fritzing

same time it can increase the delay of the R-peak detection. Thus, it is essential to strike a balance between accuracy and delay based on the specific use case. Lastly, in regard of software-related time delays, the metric frames per second (FPS) are highly relevant. The FPS rate refers to the amount of frames that can be rendered and displayed on the screen per second and is dependent on the quality of the hardware, but also many other factors, such as the overall complexity of the scene, the amount of objects in the scene, the quality of the graphical rendering, and other optimization parameters, e.g. level-of-detail (LOD) systems, occlusion culling, and dynamic batching. Normally, the FPS rate is between 30 and 60 FPS, but can be upscaled if necessary (Fig. 4).

3 Results

3.1 Accuracy in Detection of Inter-Beat-Intervals

To investigate the accuracy in detection of IBIs, IBIs of all participants for all four signals were calculated. IBIs smaller than 700 ms or larger than 1500 ms were marked as outliers and were therefore removed. An overview over descriptive statistics can be found in Table 2, and a histogram displaying all IBIs can be found in Fig. 5. Overall, all four methods perform similar in regard to means and standard deviations for the identified IBIs. The ArdMob-ECG performed similarly to the AdInstruments ECG in the number of IBIs detected, indicating that the ArdMob-ECG is a suitable tool for the assessment of R-peaks.

IBIs, interpolated heartrate time-series, and Poincare Plots for all four signals were plotted using the r-package RHRV [29] for one randomly chosen participant (Fig. 6. Plots for the other participants can be found under this link: https://github.com/wunderwald/Arduino_ECG). The results of our analysis indicate that there are negligible differences in the plots of the inter-beat-intervals (IBIs), interpolated heart rates (HRs), and Poincare plots between the ArdMob-ECG and the AdInstruments

Table 2 Descriptive statistics comparing the time difference between two consecutive R-peaks (IBIs) that were identified by running a peak-detection algorithm over the AdInstruments ECG data and manually correcting misidentified R-peaks (total R-peaks identified $N = 1405$); Maximum number of identifiable IBIs = 1401. The single not-identified IBI in ADInstruments ECG and ArdMob-ECG has been the same

Signal	Mean	SD	IBIs identified
AdInstruments ECG	944.14 ms	103.96 ms	1400
ArdMob-ECG	944.15 ms	104.06 ms	1400
FastResponseOutput	944.67 ms	104.26 ms	1389
Finger Sensor	944.60 ms	105.30 ms	1383

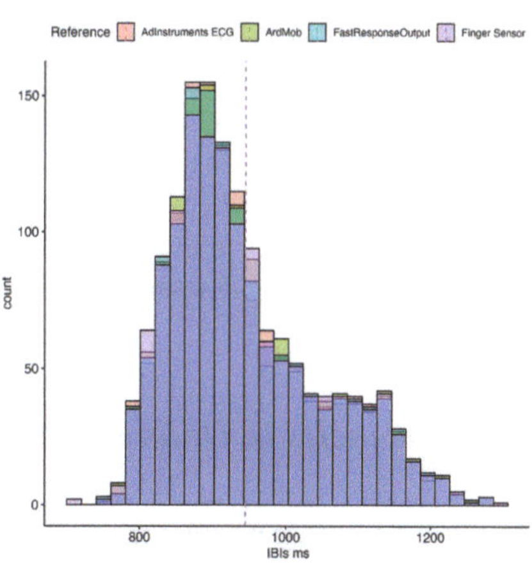

Fig. 5 Histogram of IBIs for all participants for each method

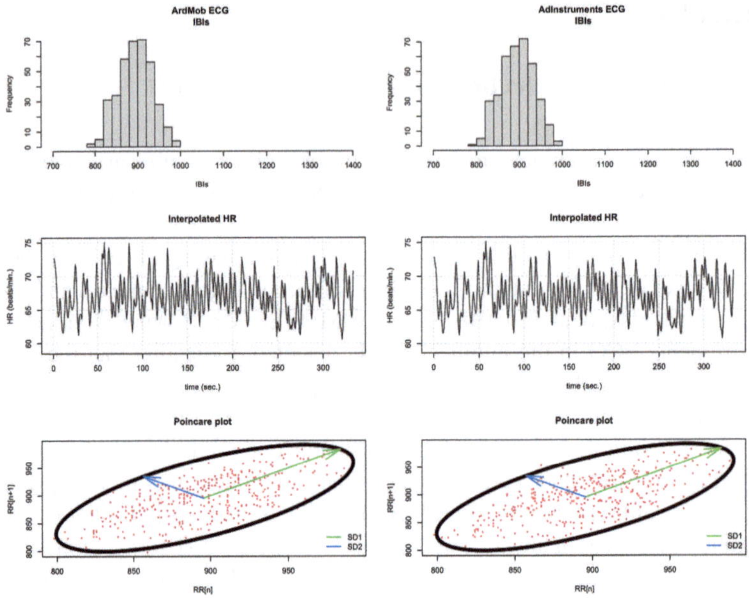

Fig. 6 Comparison of signals for one randomly selected participant. Histogram of IBIs, timeline of interpolated heartrate (HR), and Poincare Plots for the ArdMob-ECG and AdInstruments ECG signal for one participant. Equivalent plots for the other participants and the finger sensor and the fast response output can be found in the supplementary materials under this link: https://github.com/wunderwald/Arduino_ECG. A Poincaré plot is a form of a scatter plot that shows the relationship between consecutive R-R intervals. The vertical axis displays the duration of the current R-R interval, while the horizontal axis represents the duration of the previous R-R interval

ECG. These findings suggest that both devices perform similarly and are comparable in their measurement of cardiac activity. Therefore, the ArdMob-ECG is a capable low-cost, portable, and wireless device for monitoring cardiac activity.

Next, a frequency domain analysis for the ArdMob-ECG (Fig. 7) and the ADInstruments ECG of the same participant (Fig. 7) was plotted. Inspection of the normalized powerbands of heart rate variability (HRV) for both the ArdMob-ECG and AdInstruments ECG also revealed negligible differences between the devices. Additionally, a closer inspection of the high-frequency (HF), low-frequency (LF), very low frequency (VLF), ultralow frequency (ULF), and LF/HF measures showed no visible differences in the plots. These findings suggest that the ArdMob-ECG is a suitable tool for the assessment of HRV (Fig. 8).

3.2 Latencies in Online R-Peak Detection

Lastly, we assessed whether the claim of the ArdMob-ECG inventor of "(near) real-time ECG analysis" holds true. As described in their paper, the ArdMob-ECG can

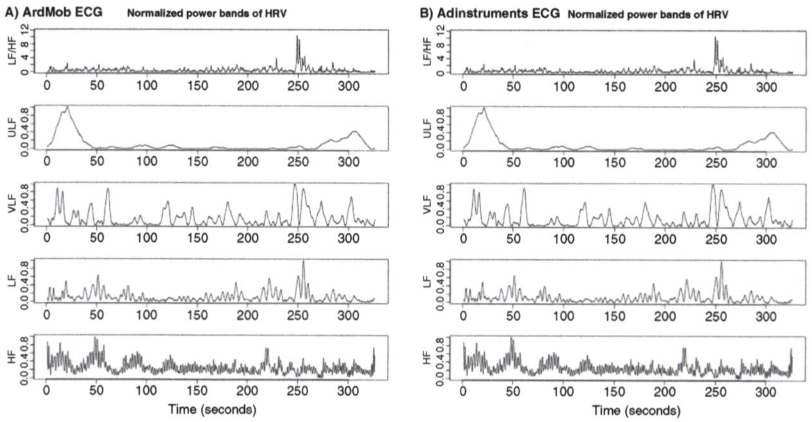

Fig. 7 Frequency domain analysis for one participant for (**a**) ArdMob-ECG and (**b**) AdInstrumentsECG. Normalized frequency domain analysis for (**a**) ArdMob-ECG and (**b**) AdInstruments ECG for one participant. Equivalent plots for the other participants can be found in the supplementary materials under this link: https://github.com/wunderwald/Arduino_ECG

Fig. 8 This figure displays the comparison of four different modules that can be used for the recording of heart-related data. It shows that the R-detect function of the ArdMob-ECG has very comparable latencies compared to state-of-the-art ECG devices. Compared to the fingersensor, all ECG devices detect the heartbeats earlier

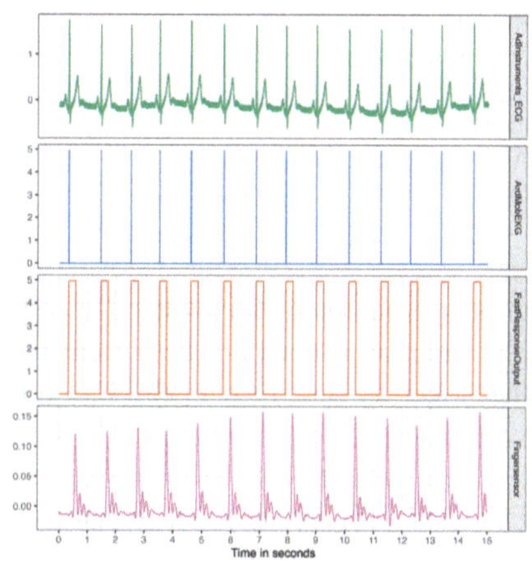

be used for displaying ones' heartbeat in near real time, e.g., in studies using virtual reality [2]. We find that on average the ArdMob-ECG detected a peak with a latency of 8 ms compared to the reference of the AdInstruments ECG (Table 3).

The other factor is the software latency. The time difference in R-peaks detected by the ArdMob-ECG and the FastResponseOutput for all participants was compared. We computed a difference score, which indicates the latency of the R-peak identified by either the ArdMob-ECG, FastResponseOutput, or Finger Sensor to R-peaks identified by the AdInstruments ECG. Descriptive statistics are reported in Table 3, and a histogram comparing the ArdMob-ECG with the FastResponseOutput can be found in Fig. 9. Negative values are possible if the R-peak is detected before the R-peak reaches its maximum, as is the case for the FastResponseOutput, which

Table 3 Descriptive statistics comparing latencies in relation to R-peaks identified by the AdInstruments ECG as a reference. Descriptive statistics comparing latencies in relation to R-peaks identified by running a peak-detection algorithm over the AdInstruments ECG data and manually correcting misidentified R-peaks (total R-peaks identified $N = 1404$)

Signal	Mean	SD	Peaks not identified	Peaks classified as outlier
ArdMob-ECG	8.02 ms	5.27 ms	1	0
FastResponseOutput	−3.51 ms	3.32 ms	9	240
Finger Sensor	281.57 ms	20.06 ms	2	239

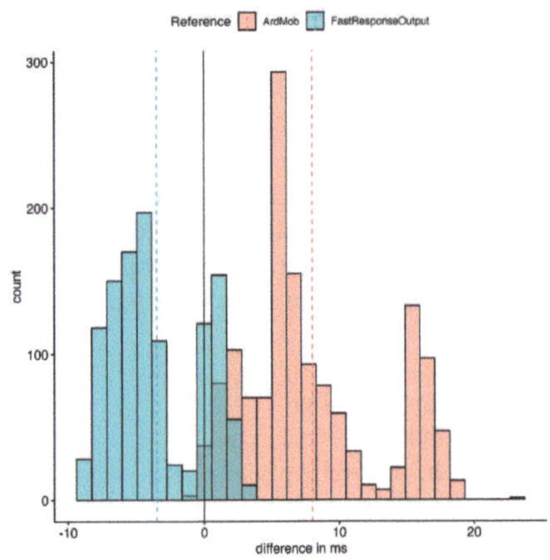

Fig. 9 Histogram comparing latencies for ArdMob-ECG (red) and FastResponseOutput (blue) in relation to R-peaks identified by running a peak-detection algorithm over the AdInstruments ECG data and manually correcting misidentified R-peaks (total R-peaks identified $N = 1405$). Thus, 0 (black line) refers to a R-peak identified by the AdInstruments ECG. Dashed lines refer to the group means

uses a predefined threshold. The relative large latency for the finger sensor is plausible, as it does not detect electrical signal, but changes in blood pressure in the finger as described in the Introduction. Latencies for the ArdMob-ECG were very close to R-peaks identified by the AdInstruments ECG as can be seen in Fig. 9. Furthermore, in contrast to FastResponseOutput and Finger Sensor, the ArdMob-ECG had a higher accuracy in detecting R-peaks. It is worth noting that the authors have defined a window size of 250 samples for the detection algorithm.

On average, the author's claim of "near real time ECG analysis" can be quantified to an average of 13 ms. Calculated by the hardware delay of 5 ms, plus 8 ms delay from the software. This value however might be different according to the specific software of use. To assess this delay empirically, we wrote a script in Unity 3D that detected when information was send by the ArdMob-ECG through the serial interface to Unity. The Unity was sending a message back to the Arduino. By taking the half of the time until the Arduino receives the message back, this can be the approximated time it takes until a detection of the heartbeat is processed by other software. The code of the script can be found under this link: https://github.com/wunderwald/Arduino_ECG. Table 4 shows an overview of the latency between the aforementioned Arduino to Unity, to Arduino setup (Fig. 4). As shown in Table 4, the median and mean rtt are around 50 ms. By taking half the rtt, the actual hardware

Table 4 Latencies of the round trip time (rtt) of the ArdMob-Unity-Arduino interface. Depicted are the mean Roundtrip times (rtts), median rtts, minimum rtts, and max rtts in this order in milliseconds. The latencies were approximated by taking half the time of the rtts. This value approximates the time until Unity 3D processed each trigger (27 triggers sent in total)

mean RTT	**median RTT**	**min RTT**	**max RTT**
50,0072	50,0000	9,0000	92,0000
mean Latency	**median Latency**	**min Latency**	**max Latency**
25,0036	25,0000	4,5000	46,0000

and software latency of the ArdMob-ECG and Unity can be approximated as around 25 milliseconds.

4 Discussion

The ArdMob-ECG was compared against three methods of assessing R-peaks using a state-of-the art ECG and pulse oximetry. We focused on two aspects—identification of IBIs and latency in detection of R-peaks. Regarding identification of IBIs, all methods used here provide a similar estimate. However, it must be noted that the R-detect function of the ArdMob-ECG and AdInstruments ECG seem to identify slightly more IBIs than FastResponseOutput and fingersensor. This is a reasonable observation given that the FastResponseOutput is not specifically designed for HR assessment, and the fingersensor is more heavily impacted by movement artifacts. Regarding latencies in R-peak detection, the ArdMob-ECG has a minimal latency with a mean of 8 ms compared to the AdInstruments ECG, which for practical applications, such as experimental paradigms using real-time feedback should have almost no impact. Nonetheless, the validation provided here can be taken into account when aiming at providing a precise feedback at certain times in the heartbeat cycle. To assess the delay in practice, a sample experiment was conducted that measured the latency of an Arduino-Unity interface that has been described in the original paper. To that end, an ArdMob-Unity-Arduino interface was developed that recorded the average round-trip time. The actual hardware and software latency was concluded to be around 25 ms on average. It is important to note, however, that the FPS rate of Unity 3D was upscaled up to an average mean of 496 frames per second. This upscaling might only be feasible in a resource-efficient Unity 3D paradigm and/or high-end hardware technology. The software-specific delays of other third-party software, e.g., Matlab or Python, that are often used will be assessed in a future analysis.

5 Conclusion

This chapter presented the validation of the ArdMob-ECG as a cost-efficient and portable alternative to classic medical ECG machines. The performance of the ArdMob-ECG is compared to other ECG and pulse-oximeter devices in regard to accuracy in detecting R-peaks, latency, and overall signal quality. Our results showed that the ArdMob-ECG performs comparable to the reference ECG. This indicates that the ArdMob-ECG can be a suitable tool for assessing R-peaks and calculating heart rate variability (HRV) and respiratory sinus arrhythmia (RSA) indices. Poincaré plots and frequency domain analyses further demonstrated the comparability of the ArdMob-ECG and the reference ECG device in measuring cardiac activity. In comparison to the pulse oximeter, the ArdMob-ECG could detect heartbeats on average 271 ms quicker and with less outliers.

The ArdMob-ECG offers several advantages over traditional medical ECG devices, including its low cost, portability, and ease of use. A three-electrode setup instead of the usual 12-electrode setup is sufficient for R-peak detection, making it more user-friendly and less cumbersome. The device can be operated on battery power, making it suitable for use in remote areas or situations where access to electricity is limited. Furthermore, the ArdMob-ECG can provide auditory feedback upon detecting a heartbeat, which can be played faster or slower compared to the actual heartbeat, which is a feature not offered by most clinical machines that can be beneficial for certain scientific paradigms.

To conclude, our validation study demonstrates that the ArdMob-ECG is a reliable and cost-effective alternative to medical ECG devices in regard to R-peak and interbeat-interval detection. It has the potential to expand access to ECG monitoring in both scientific research and clinical settings. Future work should focus on further validating the device's performance in larger and more diverse populations and settings. Additional features like WiFi- and Bluetooth transmission, as well as an inbuilt screen to monitor the heartbeat data, are currently in progress.

Acknowledgments The original paper can be accessed under https://arxiv.org/abs/2204.00513. The pictures and the basic code for this device can be found in the GitHub repository [30]. The repository is licensed under the MIT license. Many thanks to Yasmin Kim Georgie for language editing and proofreading.

Author Contributions: T.J.M.: Conceptualization, software, resources, writing-original draft & editing, visualization, and graphical design; M.W.: Review & editing, software adaptations & improvements M.T.: Editing, visualization, data acquisition, graphical design.

References

1. T. Möller, Y. Georgie, M. Voss, and L. Kaltwasser, "An Arduino based heartbeat detection device (ArdMob-ECG) for real-time ECG analysis," *2022 IEEE Signal Processing in Medicine and Biology Symposium (SPMB)*, 2022, pp. 1–3.

2. T. J. Möller, M. Voss, and L. Kaltwasser, "An Arduino based heartbeat detection device (ArdMob-ECG) for real-time ECG analysis," *arXiv preprint arXiv:2204.00513*, 2022.
3. S. N. Garfinkel, A. K. Seth, A. B. Barrett, K. Suzuki, and H. D. Critchley, "Knowing your own heart: distinguishing interoceptive accuracy from interoceptive awareness," *Biological psychology*, vol. 104, pp. 65–74, 2015.
4. K. Suzuki, S. N. Garfinkel, H. D. Critchley, and A. K. Seth, "Multisensory integration across exteroceptive and interoceptive domains modulates self-experience in the rubber-hand illusion," *Neuropsychologia*, vol. 51, no. 13, pp. 2909–2917, 2013.
5. J. Brener, X. Liu, and C. Ring, "A method of constant stimuli for examining heartbeat detection: Comparison with the Brener-Kluvitse and Whitehead methods," *Psychophysiology*, vol. 30, no. 6, pp. 657–665, 1993.
6. S. Wiens and S. N. Palmer, "Quadratic trend analysis and heartbeat detection," *Biological psychology*, vol. 58, no. 2, pp. 159–175, 2001.
7. S. Saminu, N. Özkurt, and I. A. Karaye, "Wavelet feature extraction for ECG beat classification," *2014 IEEE 6th International Conference on Adaptive Science & Technology (ICAST)*. IEEE, 2014, pp. 1–6.
8. "Analog Devices AD8232 overview," https://www.analog.com/en/products/ad8232.html#product-overview, [Online; accessed 15-December-2021].
9. R. Schandry, "Heart beat perception and emotional experience," *Psychophysiology*, vol. 18, no. 4, pp. 483–488, 1981.
10. W. E. Whitehead, V. M. Drescher, P. Heiman, and B. Blackwell, "Relation of heart rate control to heartbeat perception," *Biofeedback and Self-regulation*, vol. 2, no. 4, pp. 371–392, 1977.
11. L. K. Lum and P. W. Cheung, "Evaluation of pulse oximetry with EKG synchronization," *Proceedings of the Annual International Conference of the IEEE Engineering in Medicine and Biology Society*. IEEE, 1988, pp. 1223–1224.
12. C. Shi, M. Goodall, J. Dumville, J. Hill, G. Norman, O. Hamer, A. Clegg, C. L. Watkins, G. Georgiou, A. Hodkinson *et al.*, "The accuracy of pulse oximetry in measuring oxygen saturation by levels of skin pigmentation: a systematic review and meta-analysis," *BMC medicine*, vol. 20, no. 1, p. 267, 2022.
13. S. DeMeulenaere, "Pulse oximetry: uses and limitations," *The Journal for Nurse Practitioners*, vol. 3, no. 5, pp. 312–317, 2007.
14. A. H. Association *et al.*, *Advanced Cardiac Life Support (ACLS) Provider Manual# 80-1088*. American Heart Association, 2006.
15. J. Allen, "Photoplethysmography and its application in clinical physiological measurement," *Physiological measurement*, vol. 28, no. 3, p. R1, 2007.
16. M. Elgendi, R. Fletcher, Y. Liang, N. Howard, N. H. Lovell, D. Abbott, K. Lim, and R. Ward, "The use of photoplethysmography for assessing hypertension," *NPJ digital medicine*, vol. 2, no. 1, p. 60, 2019.
17. I. Sadek, J. Biswas, and B. Abdulrazak, "Ballistocardiogram signal processing: A review," *Health information science and systems*, vol. 7, pp. 1–23, 2019.
18. S. Mansouri, T. Alhadidi, S. Chabchoub, and R. B. Salah, "Impedance cardiography: recent applications and developments," *Biomedical Research*, vol. 29, no. 19, pp. 3542–3552, 2018.
19. W. Wang, S. Stuijk, and G. De Haan, "Exploiting spatial redundancy of image sensor for motion robust RPPG," *IEEE transactions on Biomedical Engineering*, vol. 62, no. 2, pp. 415–425, 2014.
20. S. L. Bennett, R. Goubran, and F. Knoefel, "Adaptive Eulerian video magnification methods to extract heart rate from thermal video," *2016 IEEE International Symposium on medical measurements and applications (MeMeA)*. IEEE, 2016, pp. 1–5.
21. F. Abnousi, G. Kang, J. Giacomini, A. Yeung, S. Zarafshar, N. Vesom, E. Ashley, R. Harrington, and C. Yong, "A novel noninvasive method for remote heart failure monitoring: the Eulerian video magnification applications in heart failure study (amplify)," *NPJ digital medicine*, vol. 2, no. 1, p. 80, 2019.
22. S. Oniani, S. I. Woolley, I. M. Pires, N. M. Garcia, T. Collins, S. Ledger, A. Pandyan *et al.*, "Reliability assessment of new and updated consumer-grade activity and heart rate monitors,"

SENSORDEVICES 2018: The Ninth International Conference on Sensor Device Technologies and Applications. IARIA, 2018.
23. S. Ramasamy and A. Balan, "Wearable sensors for ECG measurement: a review," *Sensor Review*, vol. 38, no. 4, pp. 412–419, 2018.
24. D. Nunan, G. R. Sandercock, and D. A. Brodie, "A quantitative systematic review of normal values for short-term heart rate variability in healthy adults," *Pacing and clinical electrophysiology*, vol. 33, no. 11, pp. 1407–1417, 2010.
25. P. S. Addison, "Wavelet transforms and the ECG: a review," *Physiological measurement*, vol. 26, no. 5, p. R155, 2005.
26. S. A. Shufni and M. Y. Mashor, "ECG signals classification based on discrete wavelet transform, time domain and frequency domain features," *2015 2nd international conference on biomedical engineering (ICoBE)*. IEEE, 2015, pp. 1–6.
27. R. Gilgen-Ammann, T. Schweizer, and T. Wyss, "RR interval signal quality of a heart rate monitor and an ECG Holter at rest and during exercise," *European journal of applied physiology*, vol. 119, no. 7, pp. 1525–1532, 2019.
28. M. P. Desai, G. Caffarena, R. Jevtic, D. G. Márquez, and A. Otero, "A low-latency, low-power FPGA implementation of ECG signal characterization using hermite polynomials," *Electronics*, vol. 10, no. 19, p. 2324, 2021.
29. L. Rodriguez-Linares, X. Vila, M. J. Lado, A. Mendez, A. Otero, and C. A. Garcia, "RHRV: Heart rate variability analysis of ECG data. R package version 4.2. 6," 2020.
30. T. J. Möller, "Pan Tompkins QRS detector and Datalogger," Dec 2021, https://github.com/nalgi/Pan-Tompkins_QRS-detector_and_Datalogger.

Index

A. Ahmed, J. Picone (eds.), *Machine Learning Applications in Medicine and Biology*, https://doi.org/10.1007/978-3-031-51893-5

MIX
Papier aus verantwortungsvollen Quellen
Paper from responsible sources
FSC® C105338

If you have any concerns about our products,
you can contact us on
ProductSafety@springernature.com

In case Publisher is established outside the EU,
the EU authorized representative is:
Springer Nature Customer Service Center GmbH
Europaplatz 3, 69115 Heidelberg, Germany

Printed by Libri Plureos GmbH
in Hamburg, Germany